IMAGES of America

PHILADELPHIA TELEVISION

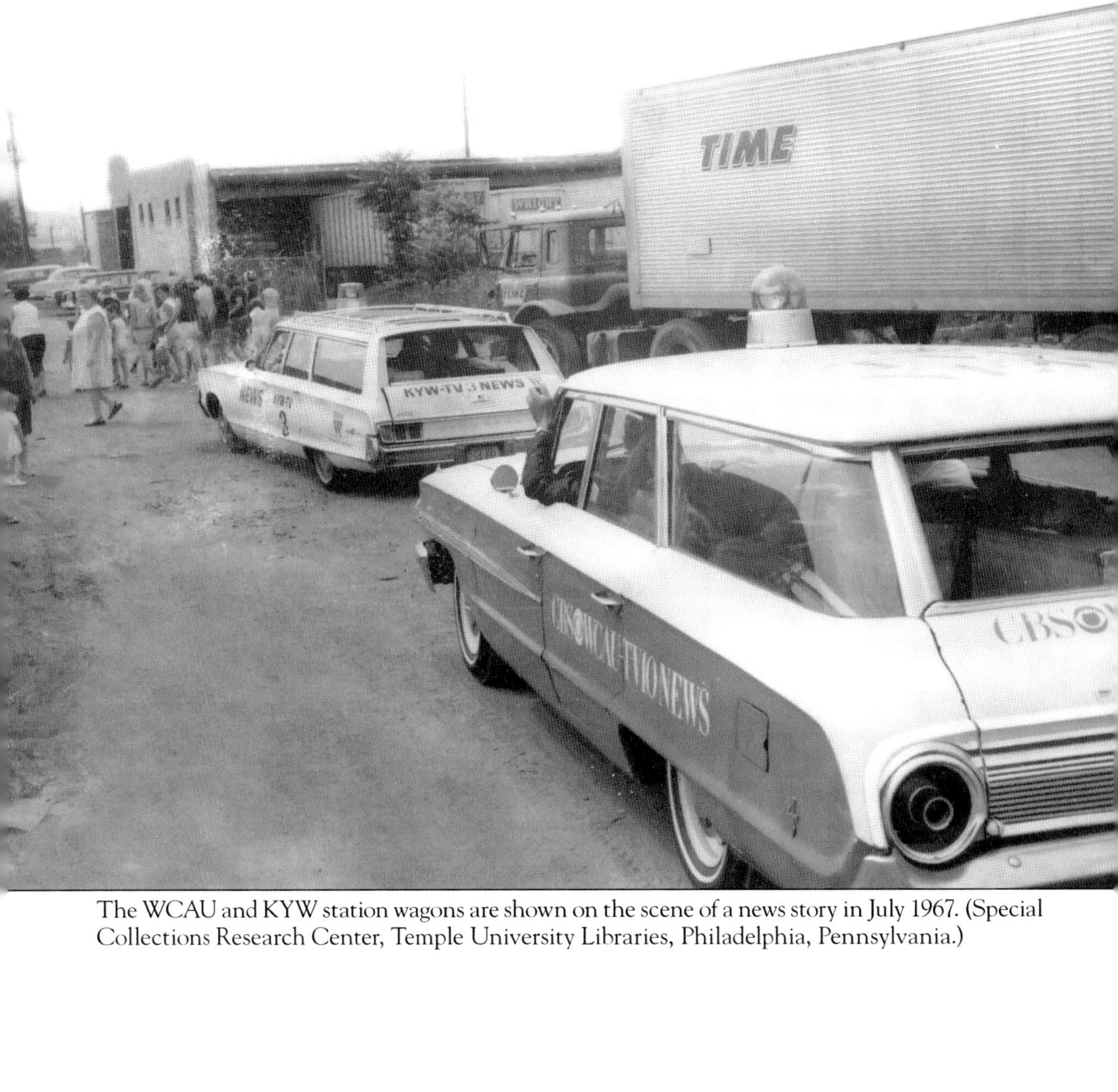

The WCAU and KYW station wagons are shown on the scene of a news story in July 1967. (Special Collections Research Center, Temple University Libraries, Philadelphia, Pennsylvania.)

On the Cover: Teenagers line up at Forty-sixth and Market Streets, the original home of *Bandstand*. The show hit the airwaves on WFIL-TV in September 1952. It originated from Studio B, and Bob Horn was the original host. He was replaced by a young disc jockey named Dick Clark in 1956. (Special Collections Research Center, Temple University Libraries, Philadelphia, Pennsylvania.)

IMAGES
of America

PHILADELPHIA TELEVISION

Bill Shull

ARCADIA
PUBLISHING

ISBN 978-1-4671-2276-4

Published by Arcadia Publishing
Charleston, South Carolina

Printed in the United States of America

Library of Congress Control Number: 2014943101

For all general information, please contact Arcadia Publishing:
Telephone 843-853-2070
Fax 843-853-0044
E-mail sales@arcadiapublishing.com
For customer service and orders:
Toll-Free 1-888-313-2665

Visit us on the Internet at www.arcadiapublishing.com

For Jill, Sue and Jenny, Judy, Joe, and Edo.

Contents

ACKNOWLEDGMENTS

When I decided to write this book, I knew I could rely on my many friends and colleagues for help. But I never imagined that so many people would go above and beyond to make this happen. People who I have worked with and people who I had never met before allowed me into their homes and offices to search personal collections.

This book would not have been possible without former WCAU anchor Tim Lake, who suggested that I take on this project. After I stopped laughing, I thought twice about doing it. Tim has written two books for Arcadia, both on the area where he grew up in New York State. His encouragement and help through the process was invaluable.

I also want to thank some big contributors: Mike Strug, Bill Baldini, and Terry Ruggles—all Channel 10 legends. I also want to thank Rex Morgan Jr., Liz Matt, Marciarose Shestack, Jane Norman, Art Ellis, Paul Gluck, Christine Zanghi, "Big Al" Meltzer, Tony Williams, Pete Kane, Jane Mihalick, Mike Nise, Kathy Gerrow, Gary Herman, Anzio Williams, Steve Levy, Cherie Bank, Jennifer Best, Bob Backman, Don Casey, and some people who I still haven't met face-to-face: Jay Lloyd, Mike Quinn, Mort Crim, Marc Howard, Mike Monsell, Vince Giannini, Rosalie Gomberg, Len Stevens, and Joanne Calabria. I also want to thank and plug two other Arcadia authors: Alan Boris, Images of America: *Philadelphia Radio*, and John Flack, Images of America: *Evesham Township*, for photographs and help with the process.

This book would have a lot of blank pages without photographs from the Special Collections Resource Center at Temple University. They do great work preserving Philadelphia history. I also want to thank the Philadelphia History Museum for opening up its archives. Director of collections Kristen Froehlich was a big help.

I want to thank the broadcast pioneers. Its members are a who's who of Philadelphia broadcasting. Without them, there would be no history to write about. The pioneers have an amazing website that chronicles the city's television and radio past. This book only touches a small fraction of what they have put together. Thanks to chairman Gerry Wilkinson for reading my text. Last, but not least, I have to mention "ROMEO" (an acronym for "retired old men eating out). This is an informal group of retired Philadelphia TV veterans. There are many women too. They meet for lunch every few months to tell stories with old friends. I am thankful you let me be part of your gatherings. A lot of what I have learned from this group has gone into this book.

Introduction

Philadelphia television can trace its roots back to a young engineer named Philo Farnsworth. He began experimenting with the new medium starting in 1931 at the Philco Company at C and Tioga Streets. He then worked in two different labs on Mermaid Lane starting in 1933. Farnsworth also experimented at the Franklin Institute. His work led to W3XE Channel 3, the city's first broadcast TV station. The station would also become the first commercial station in Philadelphia, as WPTZ. It was the first to broadcast some programs in color. Channel 3 was first owned by Philco, and was sold to Westinghouse on June 1, 1953. It became WRCV when it was purchased by NBC in 1956. Westinghouse reacquired the station in June 1965 and moved the call letters from Cleveland after swapping stations with NBC. Channel 3's first facilities were in the Philco building at C and Tioga Streets. During World War II, the station relocated to its transmitter site in Wyndmoor. In 1946, WPTZ moved into the Architect's Building at Seventeenth and Sansom Streets in Center City. The station moved into the KYW radio building at 1619 Walnut Street about a year later. In 1972, Channel 3 built a new building on Independence Mall, giving it a front-row seat for the city's Bicentennial celebration in 1976. KYW is still based in Center City, at 1555 Hamilton Street.

WCAU-TV went on the air May 23, 1948. Channel 10 was first located in the building erected for WCAU Radio (but the structure was also designed for a future TV station, according to the company) at 1622 Chestnut Street. It was owned by the *Philadelphia Bulletin* newspaper. The station's transmitter was atop the PSFS building, then the city's tallest skyscraper. There was a plan to move the entire operation there, but the *Bulletin* decided to make history instead. It constructed the first building in the world designed to be a TV/radio facility from the ground up. (WFIL has also claimed its structure at Forty-sixth and Market Streets was the first built for TV.) The new building opened in 1952 at the corner of City Avenue and Monument Road on the Montgomery County side of the street, making WCAU the first Philadelphia station to leave the city limits. WCAU was also the first affiliate of the CBS Television Network. CBS bought the station from the *Bulletin* in 1958. NBC became the owner of WCAU on September 10, 1995, as part of an affiliate swap that saw CBS acquire KYW Channel 3.

WFIL-TV Channel 6 started broadcasting on September 13, 1947. The station was owned by the *Philadelphia Inquirer* (part of publisher Walter Annenberg's media empire, which included *TV Guide*). The first Channel 6 studios were on the 11th and 18th floors of the Widener Building at Broad and Chestnut Streets in Center City. In 1947, WFIL moved to a brand new building at Forty-sixth and Market Streets in West Philadelphia. The building was the first ever built to be a TV station (WFIL AM/FM radio moved in later). The building became famous as the first home of *Bandstand*, which went on the air in September 1952. In 1964, Channel 6 moved to a new, round building at 4100 City Avenue, directly across from Channel 10. (It is on the Philadelphia side, so Channel 6 employees have always paid city wage tax.) Capital Cities Broadcasting purchased Channel 6 in 1970. On April 27, 1971, WFIL changed call letters to WPVI. But arguably the biggest event in Channel 6 history happened in April 1970. That's when the station launched *Action News*, a fast-paced newscast anchored by a young journalist named Larry Kane. Viewers immediately connected with the new format, and *Action News* became a dynasty that has lasted more than 45 years. WPVI is currently owned by ABC/Disney and is still at 4100 City Avenue, but in a new building constructed for the digital age that opened in 2009.

WHYY has been a nomad for much of its existence. Its first studios were on Chestnut Street in Center City, in the former home of WCAU-TV. In 1963, the station moved to the old WFIL-TV building at Forty-sixth and Market Streets. It is currently located in the former Living History Museum at Sixth and Race Streets, just up the street from Independence Hall. WHYY has always been the city's educational broadcasting station, originally owned by the School District of Philadelphia. It started life on September 2, 1957, on UHF Channel 35. But at the time, not many TV sets could pick up UHF signals. In 1963, WHYY reached a deal with the FCC that allowed it to move to Channel 12. Channel 35 remained an educational station with the call letters WUHY (currently WYBE). WHYY was able to get the better VHF dial position when the owners of WDEL-TV Channel 7 in Wilmington, Delaware, gave up their FCC license. The station was only on the air from 1949 to 1951. WDEL, an NBC and DuMont affiliate, was the only network station in Delaware until WRDE in Rehoboth Beach went on the air as an NBC affiliate in June 2014. The switch greatly improved WHYY's reach and enabled Channel 12 to become a regional Public Broadcasting Service station, serving Philadelphia and all of Delaware.

Philadelphia television saw a major expansion in 1965 when three new UHF stations hit the airwaves: WPHL Channel 17, WIBF-TV Channel 29, and WKBS Channel 48.

Channel 29 signed on the air May 16, 1965, technically the city's second commercial UHF station. It was owned by brothers William, Irwin, and Benjamin Fox (WIBF). The first studios were located alongside WIBF radio in the Benson East apartment building on Old York Road in Jenkintown. Channel 29 was the second station located outside the city limits. The Fox brothers sold the station to Cincinnati-based Taft Broadcasting in 1969. The station then moved to its current building at Fourth and Markets Streets in the heart of Old City, about two blocks from Independence Hall. In 1986, Channel 29 became an affiliate of the fledgling Fox Network. TVX Broadcasting acquired Channel 29 in 1987 and changed the call letters to WTXF. Paramount owned the station from 1991 until 1995, when Fox Broadcasting bought the station (Fox Broadcasting is no relation to the original owners, the Fox brothers). Channel 29 has been home to the Philadelphia Flyers, 76ers, and Phillies at various times in its history.

Channel 17 first hit the airwaves as WPCA from 1960 to 1962, the city's first commercial UHF station. The channel went dark until new owners Len Stevens and Aaron Katz flipped the switch on WPHL Channel 17 on September 17, 1965. Channel 17 quickly became a favorite of younger viewers for shows like *Colorful Cartoon Club* (which featured the "Peanut Gallery" and was hosted by Philadelphia TV veteran Bill "Wee Willie" Webber), Doctor Shock's *Horror Theater*, and the afternoon dance party show *Dancin' On Air*. WPHL was also the home of the Philadelphia Phillies for most of the last 40 years (the games moved to Channel 29 for a brief time in the 1980s). Tribune Broadcasting bought the station in 1991. It became an affiliate of MyNetworkTV in 2006.

WKBS-TV Channel 48 also went on the air in 1965. It was licensed to Burlington, New Jersey, but was part of the Philadelphia market. Channel 48 was owned by California-based Kaiser Broadcasting until 1977, when Chicago's Field Communications took over. Channel 48 was also popular with afterschool viewers for its cartoons and reruns. Field put the station up for sale in 1983. No buyer was found, and WKBS went off the air on August 30, ending 18 years of broadcasting. Channel 48 is back on the air as of 2014 as WGTW. Another UHF station to make an impact in Philadelphia is Vineland, New Jersey–based WPSG Channel 57. Al Tedesco put the station on the air in 1981. It is now a CW Network affiliate owned by CBS. WNJS Channel 23 (NJTV), WPPX Channel 61 (Ion), WWSI Channel 62 (Telemundo), and WUVP Channel 65 (Univision) are also broadcast UHF stations in Philadelphia.

One

WPTZ/WRCV/KYW CHANNEL 3

Channel 3's first call letters were WPTZ, when it was owned by TV/radio manufacturer Philco. NBC bought the station in 1957 and changed the call letters to WRCV. Westinghouse took over in 1965 and moved the call letters KYW from its station in Cleveland. KYW has been a CBS owned-and-operated station since September 1995. The switch was part of a larger shuffle that saw NBC move to WCAU. KYW's *Eyewitness News* has had many format changes over the decades, from the traditional to the wild "direct connection disco news" of the early 1980s. Reporters Trudy Haynes and Malcolm Poindexter were the first African American TV reporters in Philadelphia. Vince Leonard was the station's anchorman from 1958 to 1980. Leonard's coanchor, Marciarose Shestak, was the city's first female anchor. Mort Crim anchored *Newswatch 5:30* with Jessica Savitch, sportscaster "Big Al" Meltzer, and weatherman Bill Kuster during KYW's glory days in the late 1960s and 1970s. Larry Mendte helped turn the station's ratings around when he arrived from WCAU in 2003. Many household names have worked at Channel 3 early or late in their careers: comedian Ernie Kovacs, Pete Boyle (father of actor Peter Boyle), David Brenner (a producer who went on be an A-list comedian), and news anchors Tom Snyder (host of NBC's *Tomorrow Show*), Jessica Savitch (anchored *NBC Nightly News* weekends until her death in 1985), Steve Bell (former news reader at *Good Morning America*), reporter Andrea Mitchell (NBC News), Maury Povich (host of KYW's *People are Talking* before the syndicated programs *A Current Affair* and *The Maury Povich Show*), and weatherman Steve Baskerville (who went on to *CBS This Morning*).

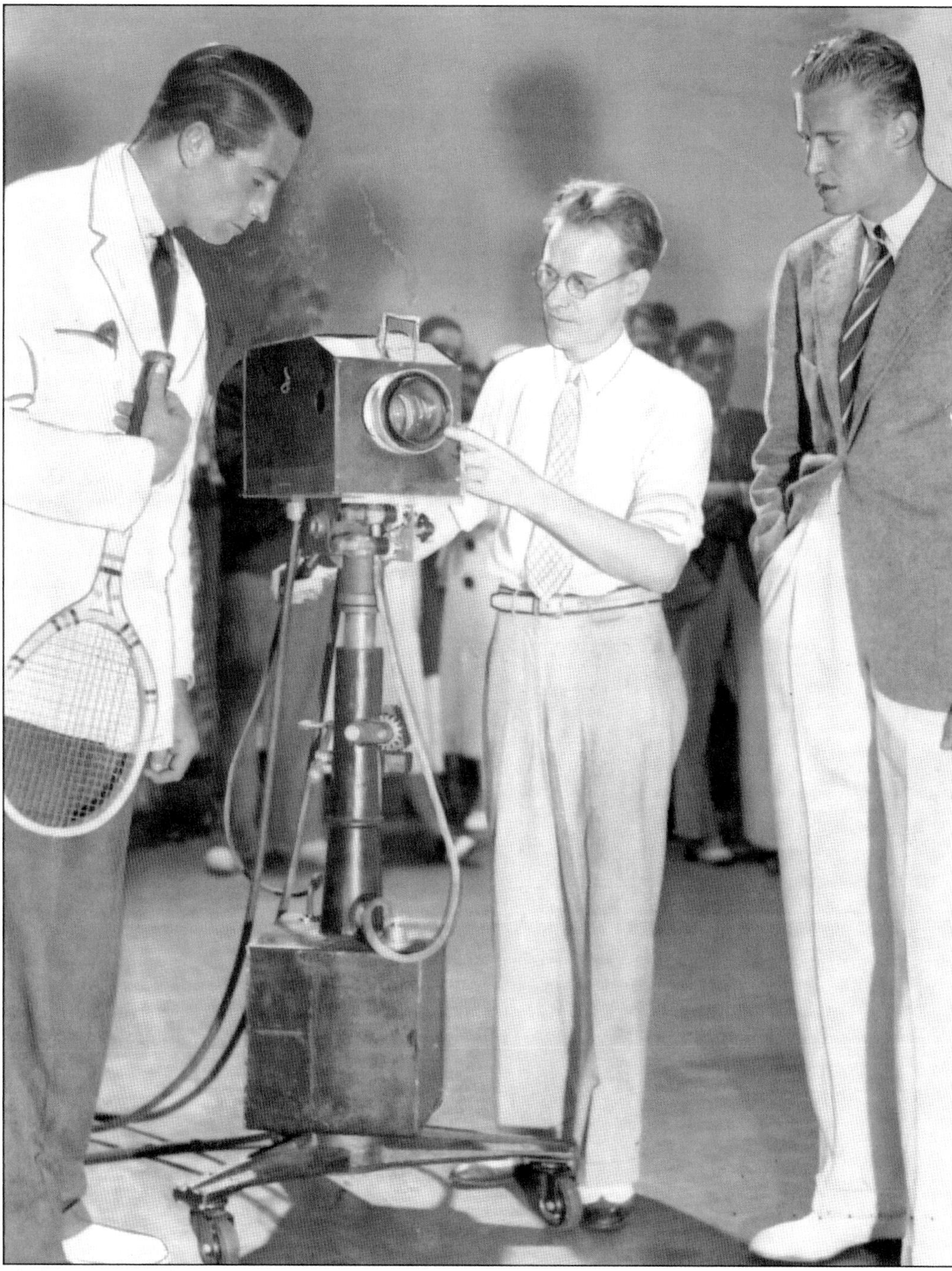

Philo Farnsworth (center) demonstrates his new invention—television—to members of the University Club in Center City in 1934. Farnsworth's experiments led to the city's first TV station, W3XE Channel 3, which began broadcasting as an experimental station in 1932. (Special Collections Research Center, Temple University Libraries, Philadelphia, Pennsylvania.)

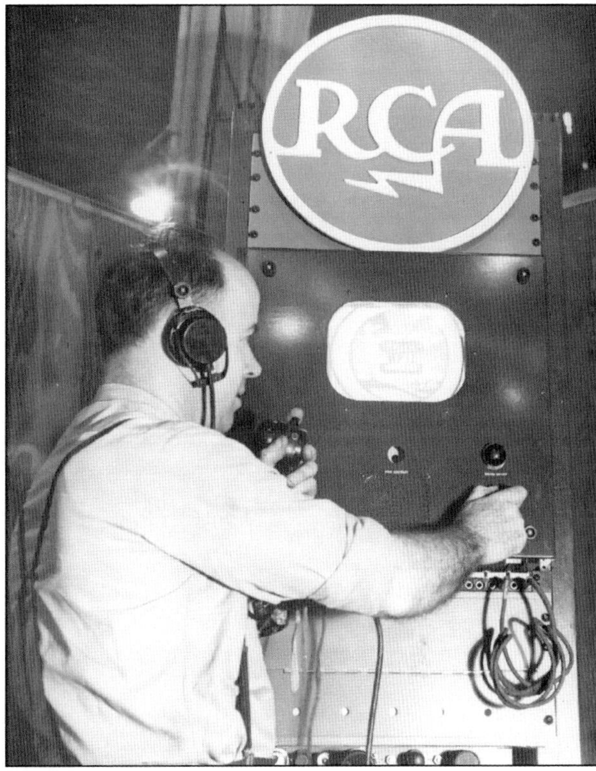

A technician monitors the broadcast in the WPTZ control room in June 1949. Channel 3 used those call letters until 1956, when it was sold to NBC. The NBC affiliate in Plattsburg, New York, picked up the call letters the same year. (Special Collections Research Center, Temple University Libraries, Philadelphia, Pennsylvania.)

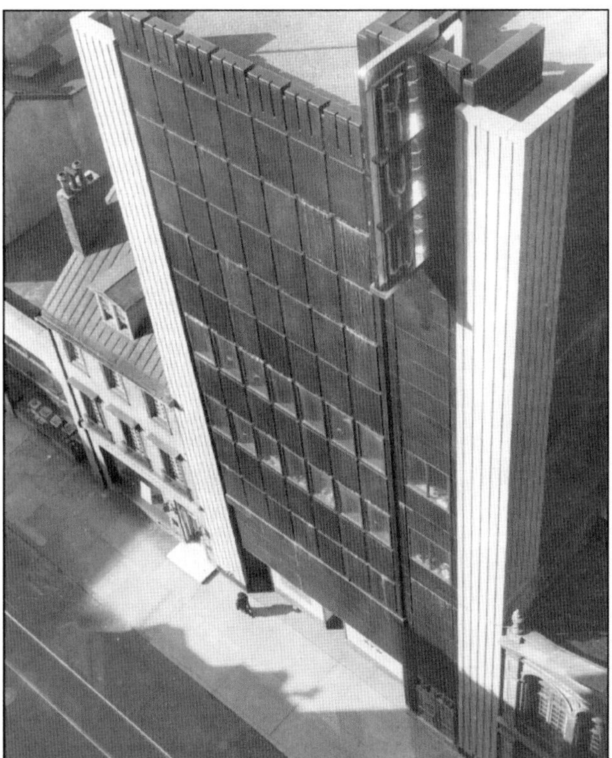

The building at 1619 Walnut Street was built for KYW radio. It cost $600,000 and opened on May 19, 1938. WPTZ moved there in 1947 from studios in the Architect's Building at Seventeenth and Sansom Streets. (Special Collections Research Center, Temple University Libraries, Philadelphia, Pennsylvania.)

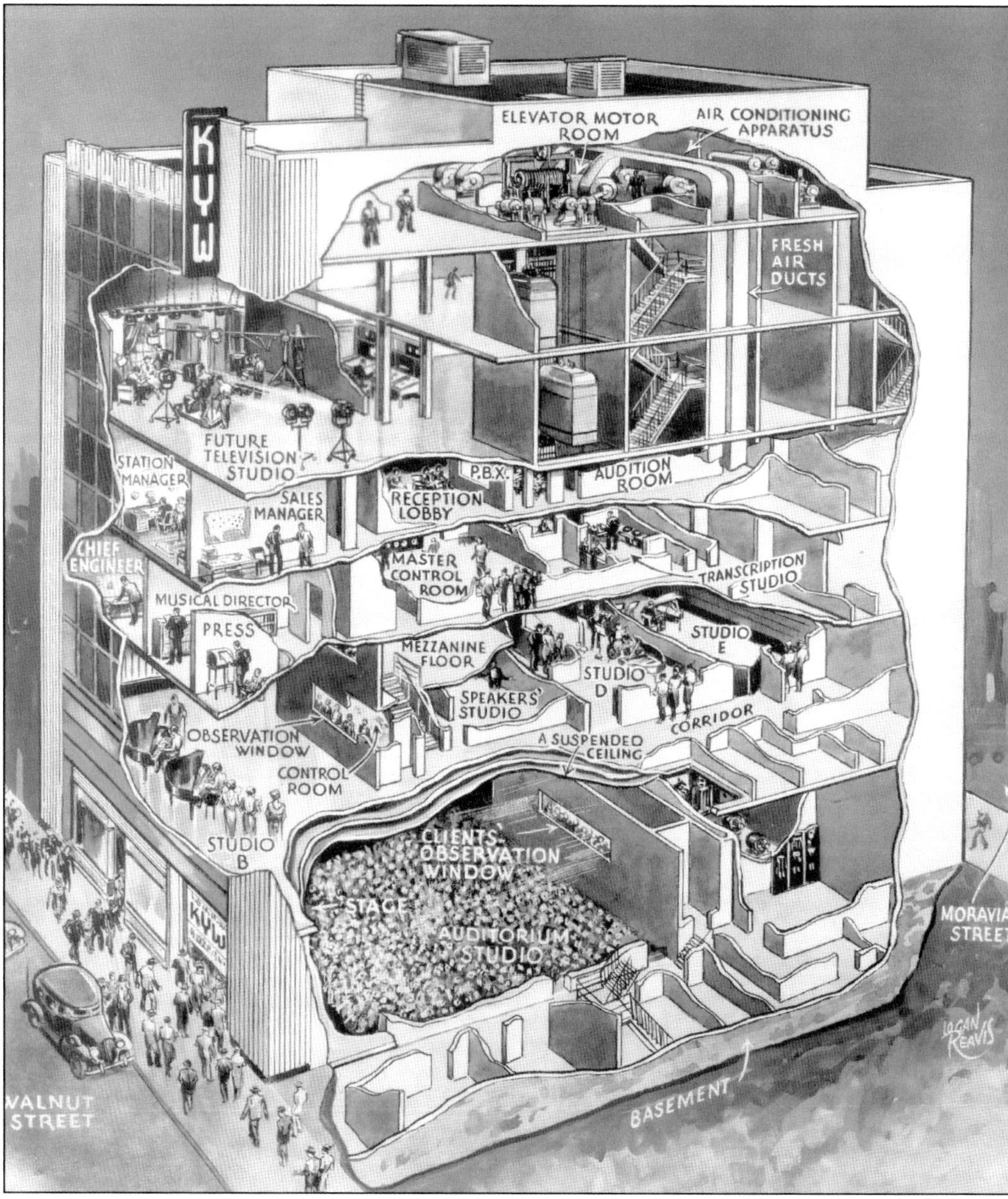

In this 1938 cutaway drawing of the 1619 Walnut building, one can see that KYW was already planning for a future TV studio on one of the upper floors. *The Mike Douglas Show* originated from the 140-seat, basement auditorium when it first moved from Cleveland in August 1965. (Special Collections Research Center, Temple University Libraries, Philadelphia, Pennsylvania.)

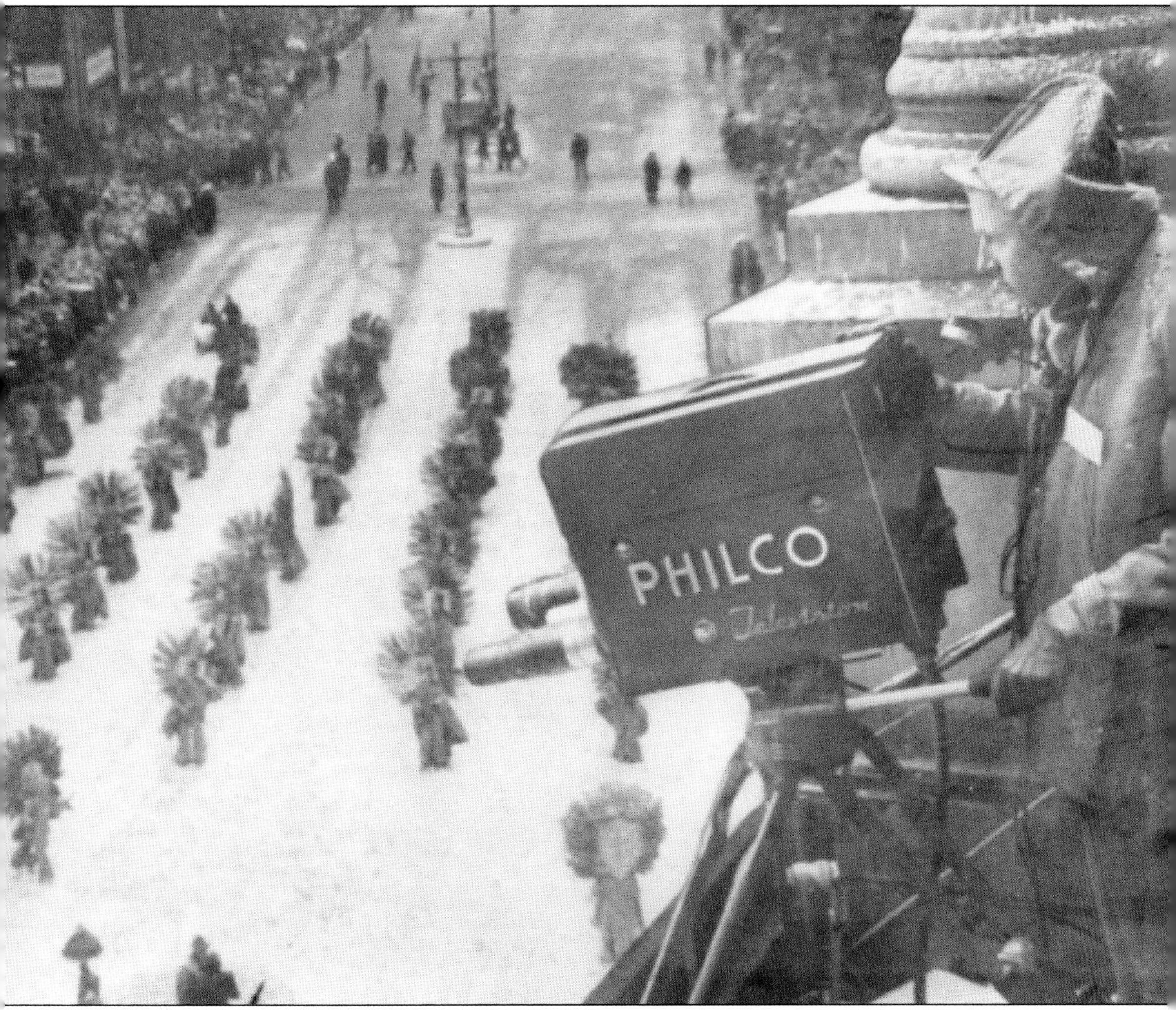

Channel 3 was the first station to broadcast Philadelphia's Mummer's Parade, January 1, 1947. This photograph shows cameraman Harold Pennypacker high above the annual strut up Broad Street. (Special Collections Research Center, Temple University Libraries, Philadelphia, Pennsylvania.)

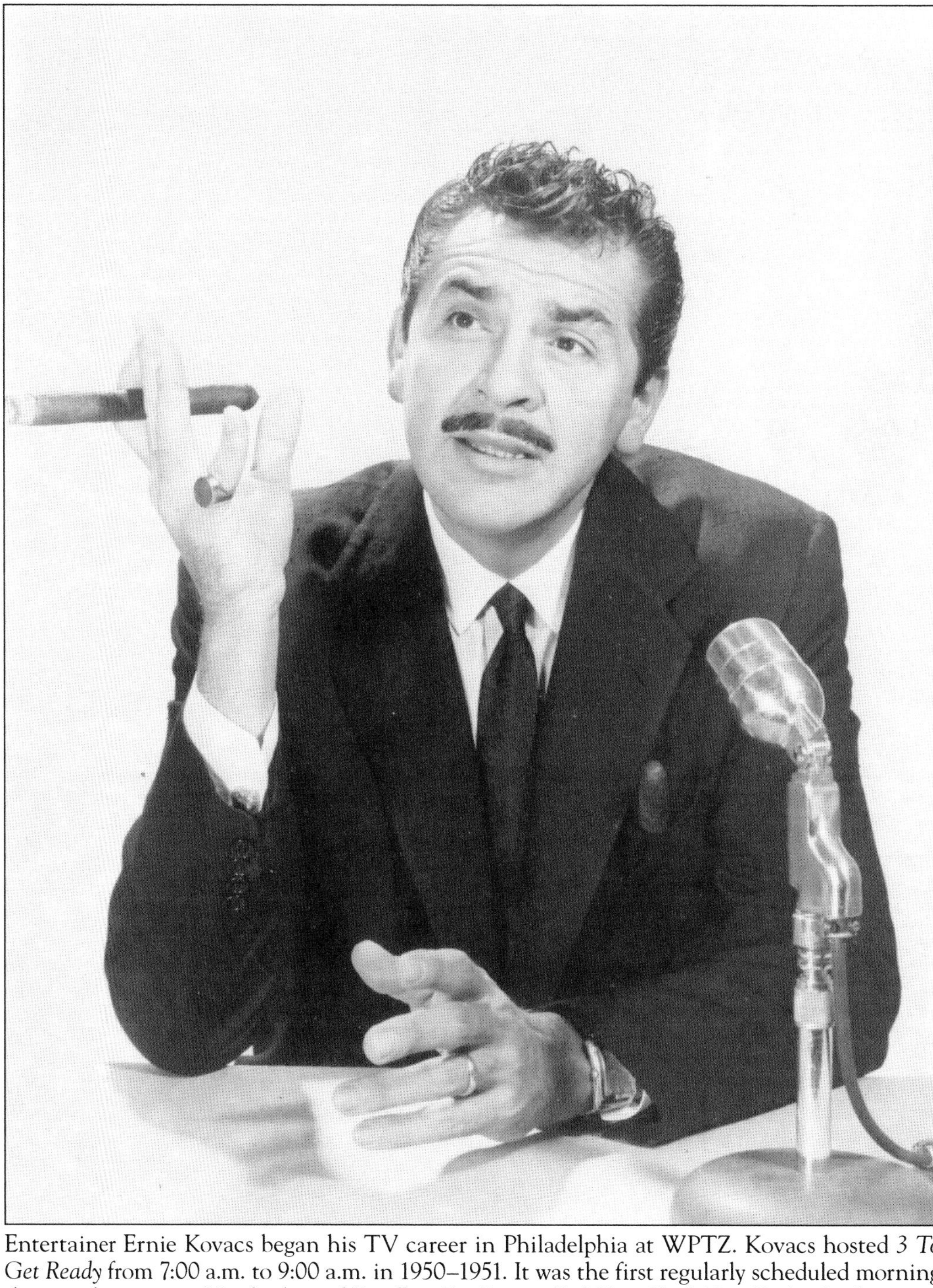

Entertainer Ernie Kovacs began his TV career in Philadelphia at WPTZ. Kovacs hosted *3 To Get Ready* from 7:00 a.m. to 9:00 a.m. in 1950–1951. It was the first regularly scheduled morning show in the country. He also hosted *Deadline For Dinner, Pick Your Ideal, Now You're Cooking*, and *Kovacs on the Corner.* Kovacs moved his shows to CBS in New York in 1952. He also had several shows on ABC and NBC. Kovacs was killed in a car accident in Los Angeles in 1962. (Special Collections Research Center, Temple University Libraries, Philadelphia, Pennsylvania.)

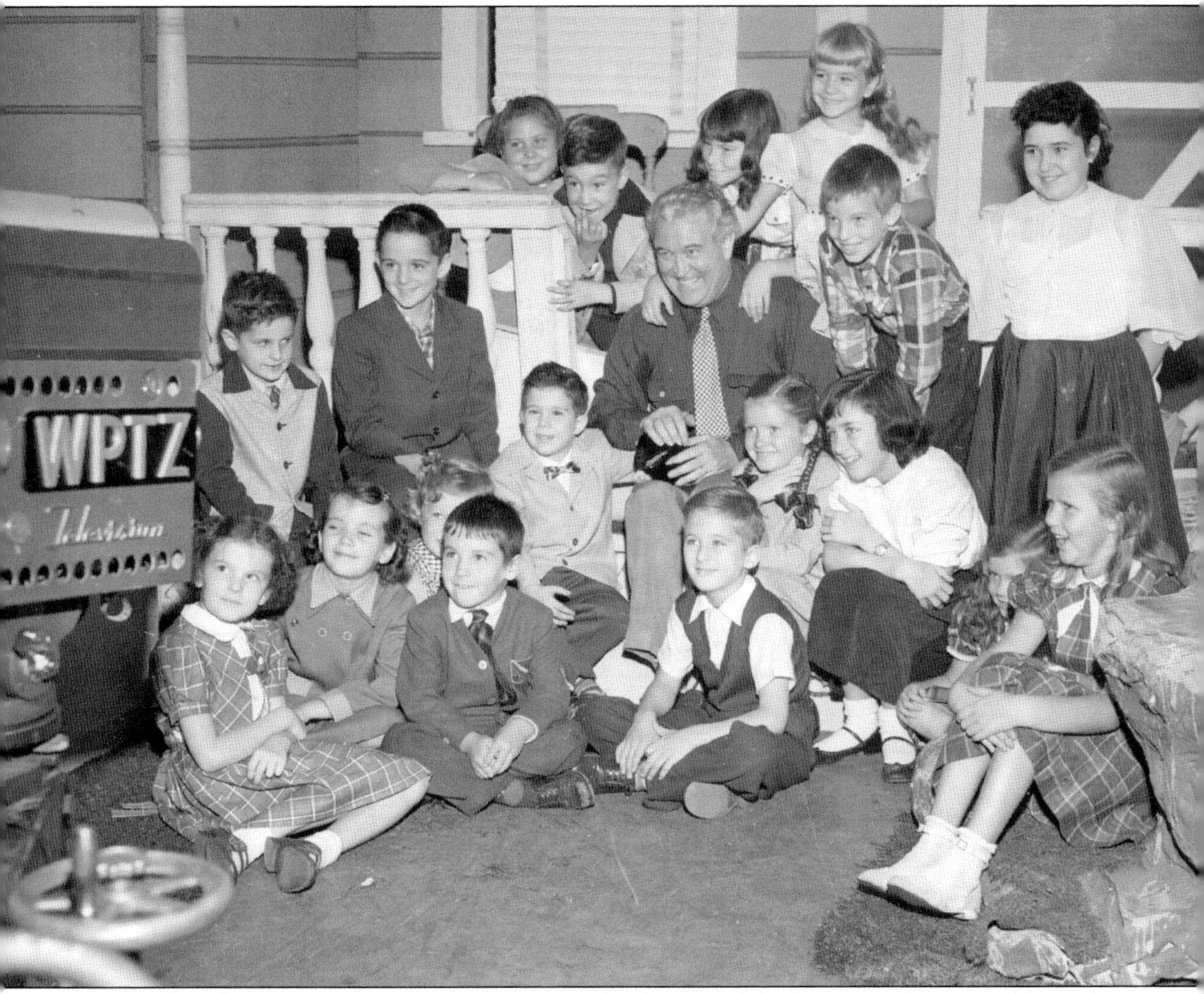

Pete Boyle hosted several children's shows on Philadelphia TV. He began at WPTZ in 1950 with a 30-minute show called *Lunch With Uncle Pete*, which ran weekdays at 12:15 p.m. Boyle was also *Chuck Wagon Pete* from 1950 to 1956. Boyle is the father of actor Peter Boyle. (Special Collections Research Center, Temple University Libraries, Philadelphia, Pennsylvania.)

Pictured: 1. **Welles Hangen,** NBC News, New Delhi; 2. **Ray Scherer,** The President On the Go; 3. **Harry K. Smith,** Weekend News, Philadelphia; 4. **Wilson Hall,** NBC News, Rio de Janeiro; 5. **Joseph C. Harsch,** NBC News, London; 6. **David Brinkley,** NBC News, Washington; 7. **Cecil Brown,** NBC News, Tokyo; 8. **Vince Leonard,** News, Philadelphia; 9. **Wally Kinnan,** Weather, Philadelphia; 10. **Sander Vanocur,** The White House, Washington; 11. **Chet Huntley,** NBC News, New York; 12. **Frank McGee,** Special Assignment; 13. **Roy Allred,** Weekend Weather, Philadelphia; 14. **Frank Bourgholtzer,** NBC News, Moscow; 15. **Irving R. Levine,** NBC News, Rome; 16. **John Chancellor,** "Today", New York. Others on NBC's crack news team, not pictured: John Rich, Europe; James Robinson, Hong Kong; Piers Anderton, Berlin; Lief Eid, Ottawa; Pauline Frederick, United Nations; Bernard Frizell, Paris; Frank Blair, "Today"; Merrill Mueller, New York; Robert McCormick, Robert Abernethy, Martin Agronsky, Herb Kaplow, Washington; Leon Pearson, New York. On the local scene, not pictured: Bob Bradley, Herb Dudnick, Dave Burrington, Joel Albert, Irv Margolis.

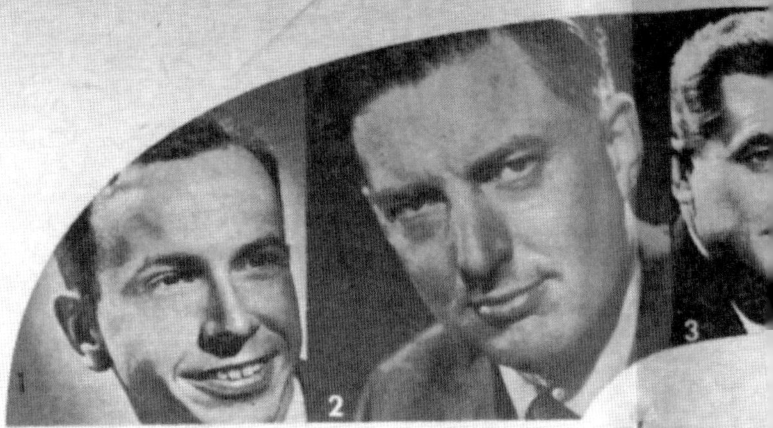

around the city
around the world

WRCV-TV
FIRST IN NEWS
&
WEATHER
around the clock

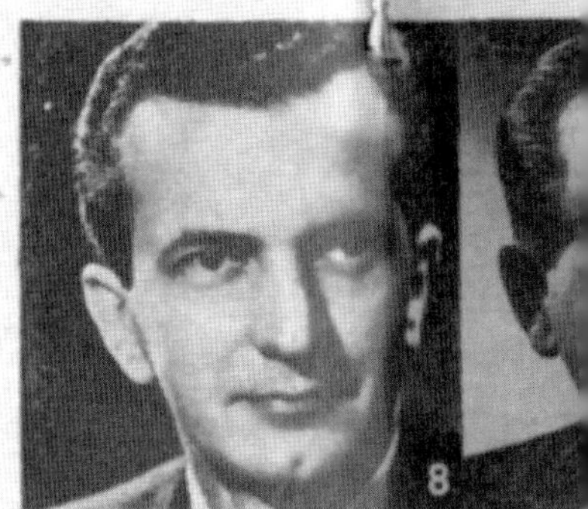

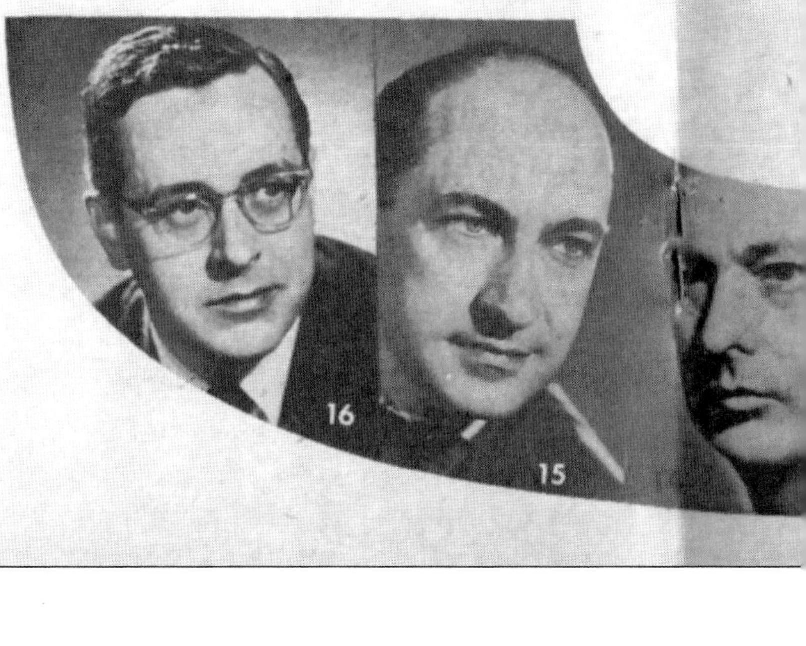

WRCV-TV
NBC in PHILADELPHIA
3

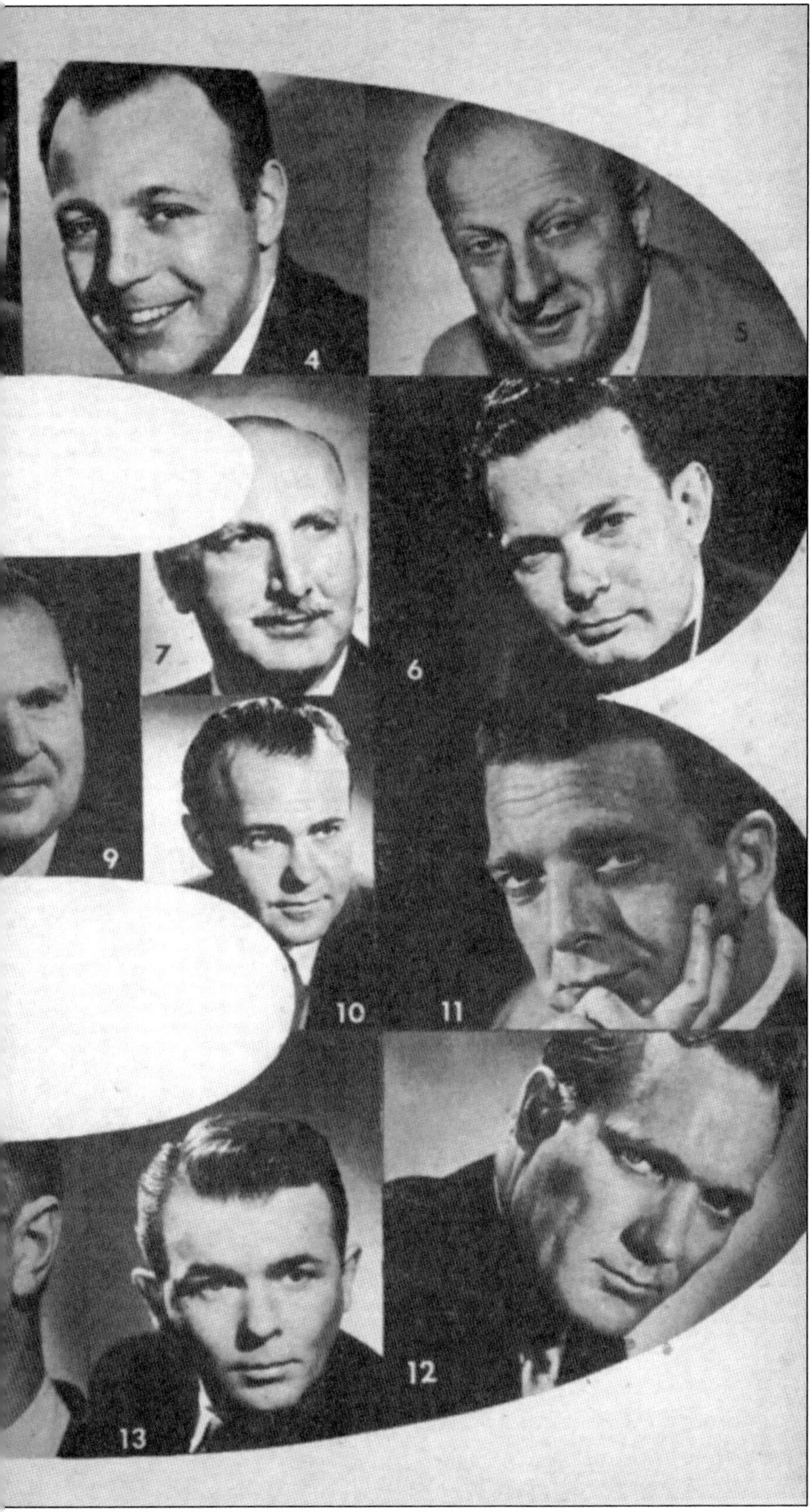

This promo from April 1962 shows a mix of WRCV and NBC News journalists; among them are anchor Vince Leonard and weatherman Wally Kinnan. Leonard began anchoring at Channel 3 in 1958. WPTZ became WRCV in 1956 after NBC bought the station from Westinghouse. (KYW-TV.)

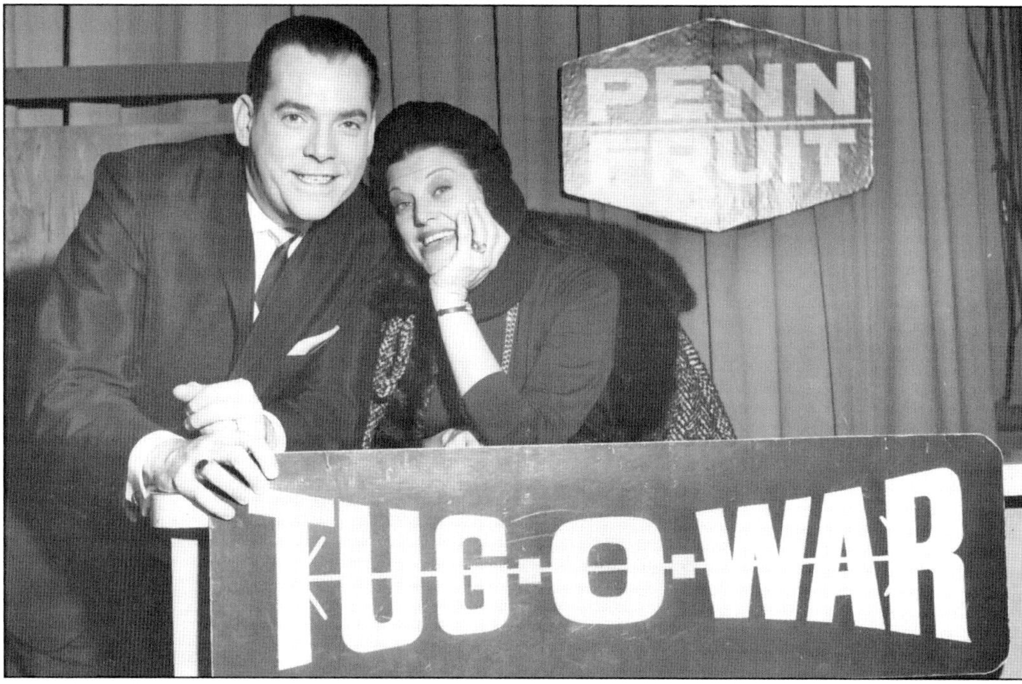

Bill Webber hosted many shows on several stations in Philadelphia. The game show *Tug-O-War* debuted in 1963 on KYW. Webber is shown here with actress Kay Ballard. (Bill Webber Jr.)

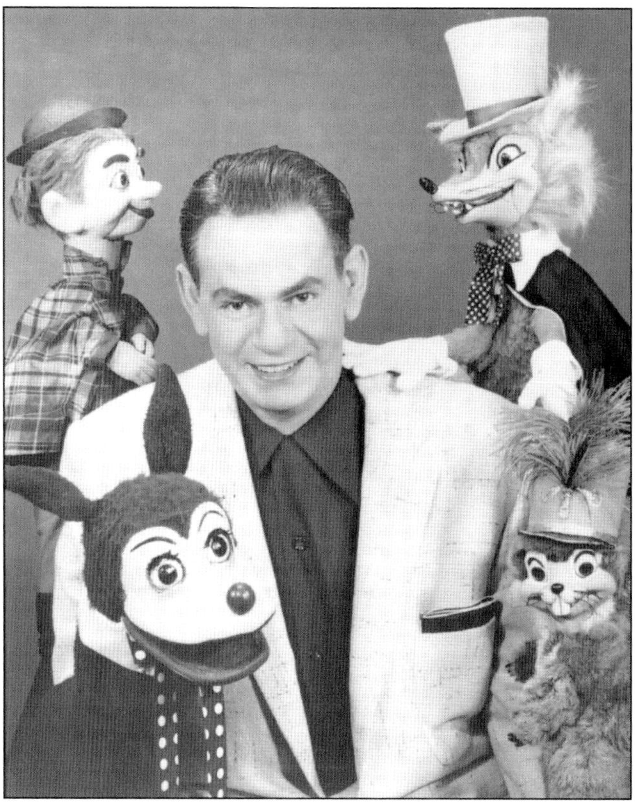

Lee Dexter was an Australian puppet master and ventriloquist. He began on WPTZ in 1950 creating puppets for Pete Boyle's children's show. He landed his own show *Bertie The Bunyip* in 1954. Dexter is seen here along with Bertie (at lower left), Fussy, Sir Guy D Guy, and Snooper the Squirrel. (Special Collections Research Center, Temple University Libraries, Philadelphia, Pennsylvania.)

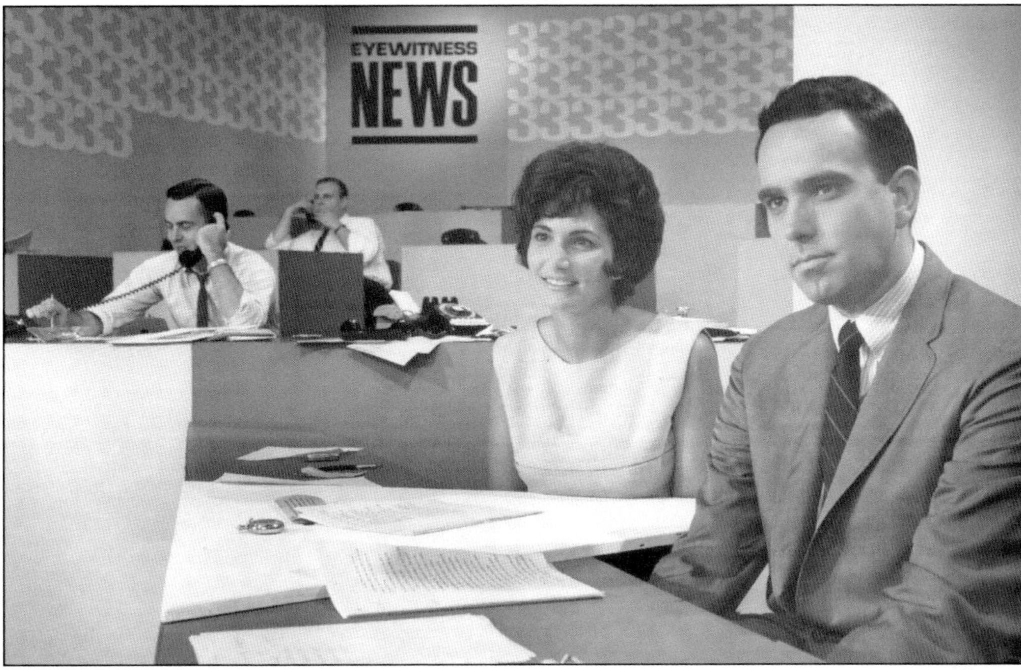

Marciarose Shestak was Philadelphia's first female TV anchor. She began at the station in 1960, hosting a talk show called *Speak Up*. Marciarose (as she was known to everyone) started coanchoring Channel 3's *Eyewitness News at Noon* with Tom Snyder in 1965. The newscast was first in its time slot. Snyder left KYW in 1970 to host NBC's *Tomorrow Show* out of Los Angeles. (KYW-TV.)

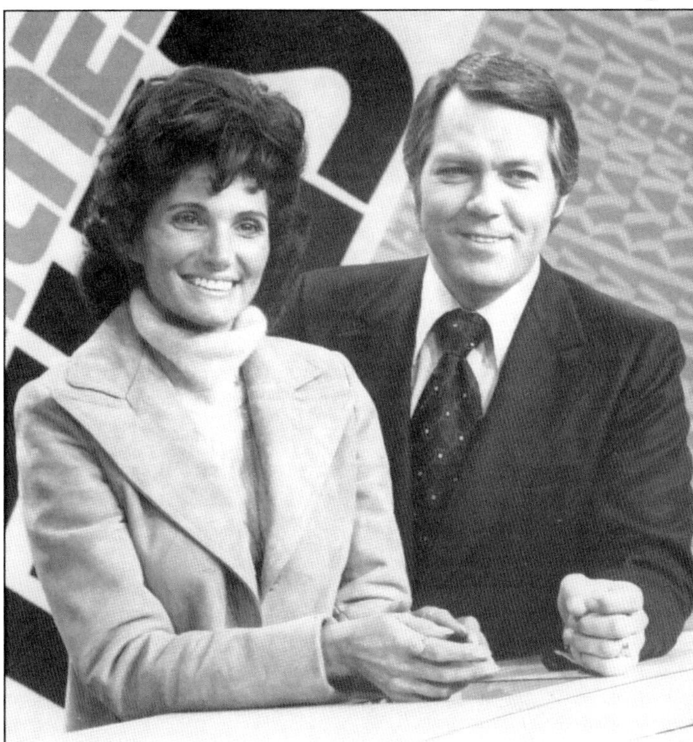

In 1973, Shestak coanchored *Eyewitness News* at 5:30 p.m. with Mort Crim. She also became the nation's first prime-time female anchor when she anchored *Eyewitness News* at 7:00 p.m. Shestak left for NBC in New York in 1975. (KYW-TV.)

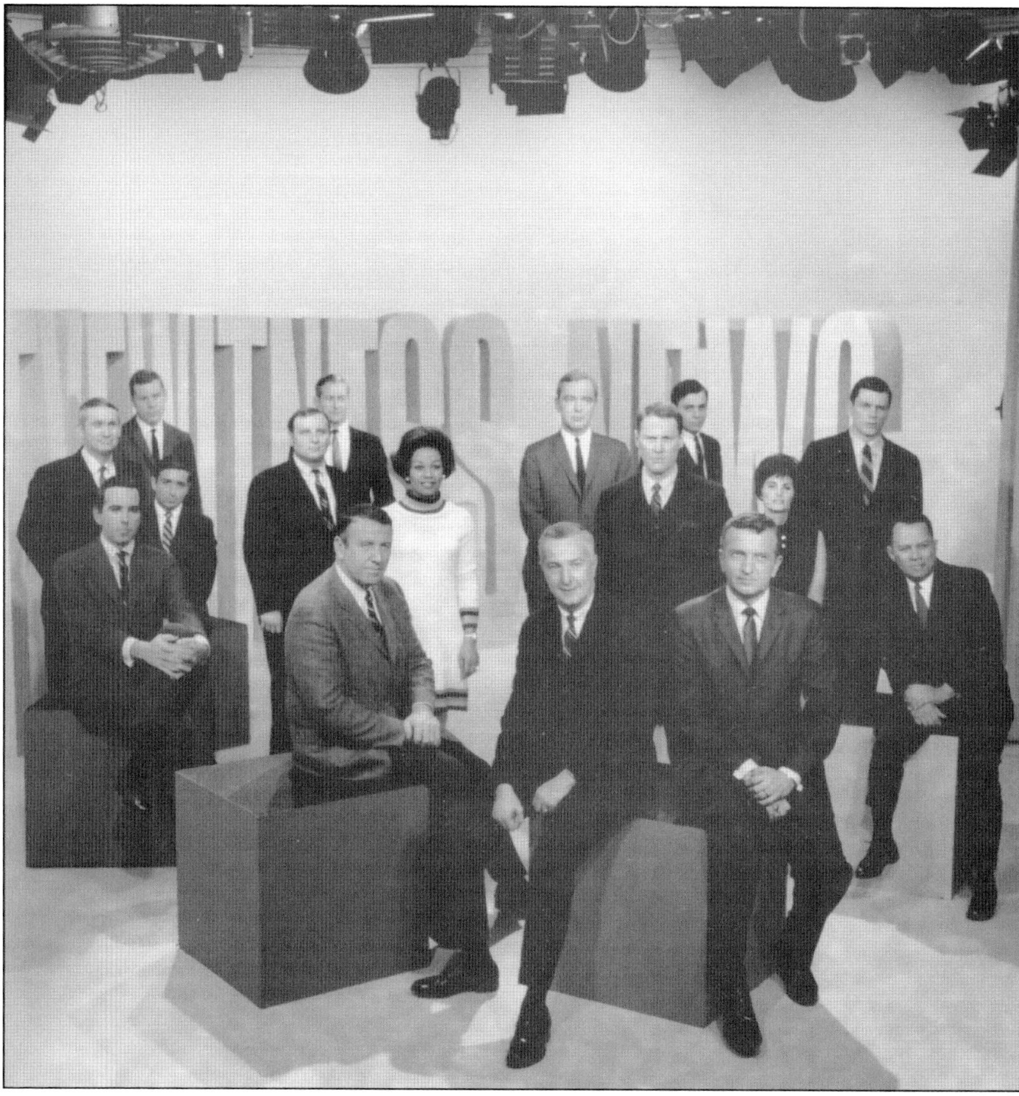

The *Eyewitness News* team is pictured in 1965. Main anchor Vince Leonard is seated in the front row on the right. Also in the image are legends Tom Snyder, Marciarose Shestak, Trudy Haynes, and Malcolm Poindexter. The format catapulted KYW to the top of the ratings ahead of WCAU and WFIL. (KYW-TV.)

Big Al Meltzer was the biggest thing in local TV sports, except when he was standing next to former Philadelphia Warriors center Wilt Chamberlin. Meltzer worked for KYW after working at WPHL. (Al Meltzer)

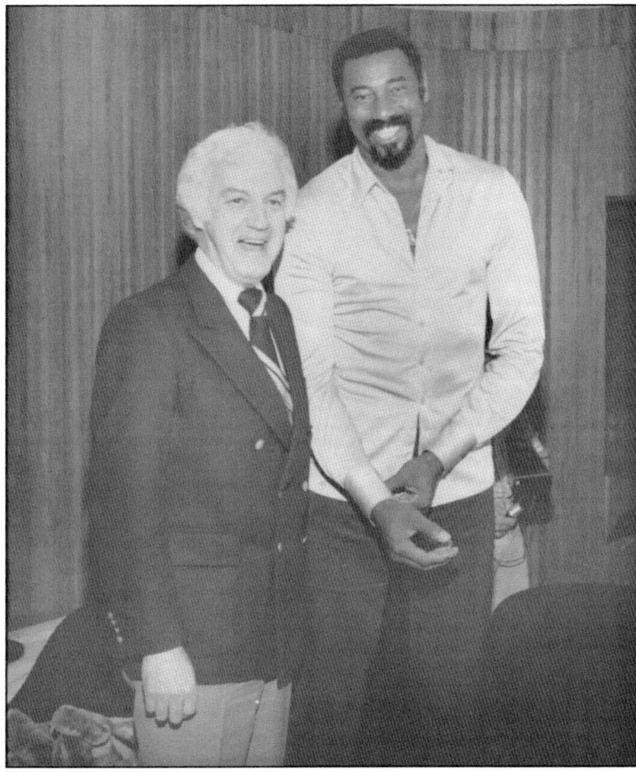

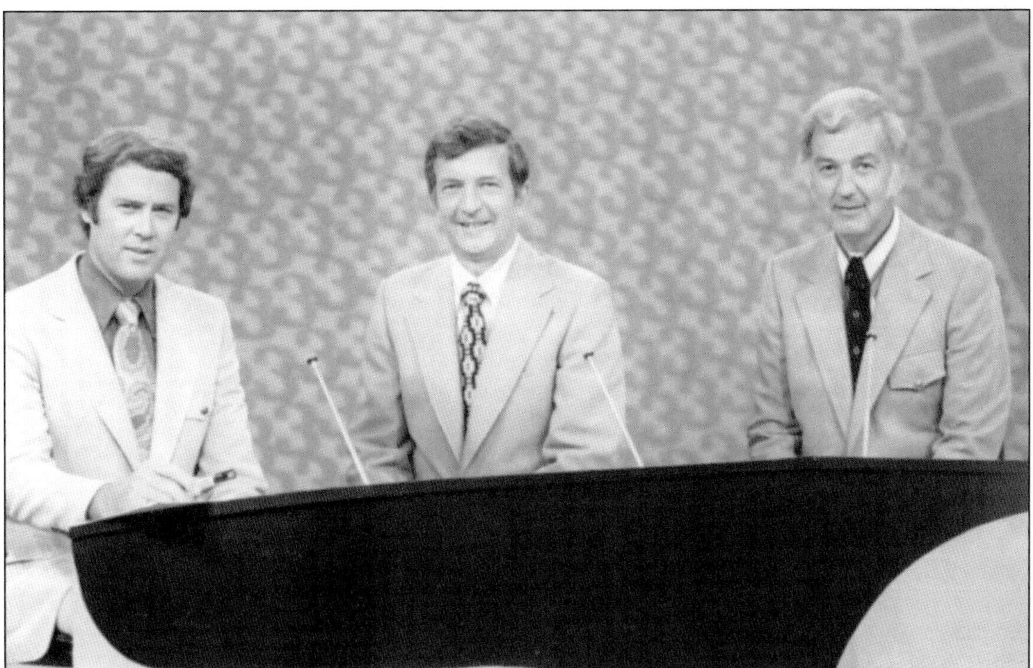

Channel 3's news team in 1970 was made up of, from left to right, meteorologist Bill Kuster, anchor Vince Leonard, and sports director Big Al Meltzer. The *Eyewitness News* format was created by news director Al Primo and copied by stations all over the country. (KYW-TV.)

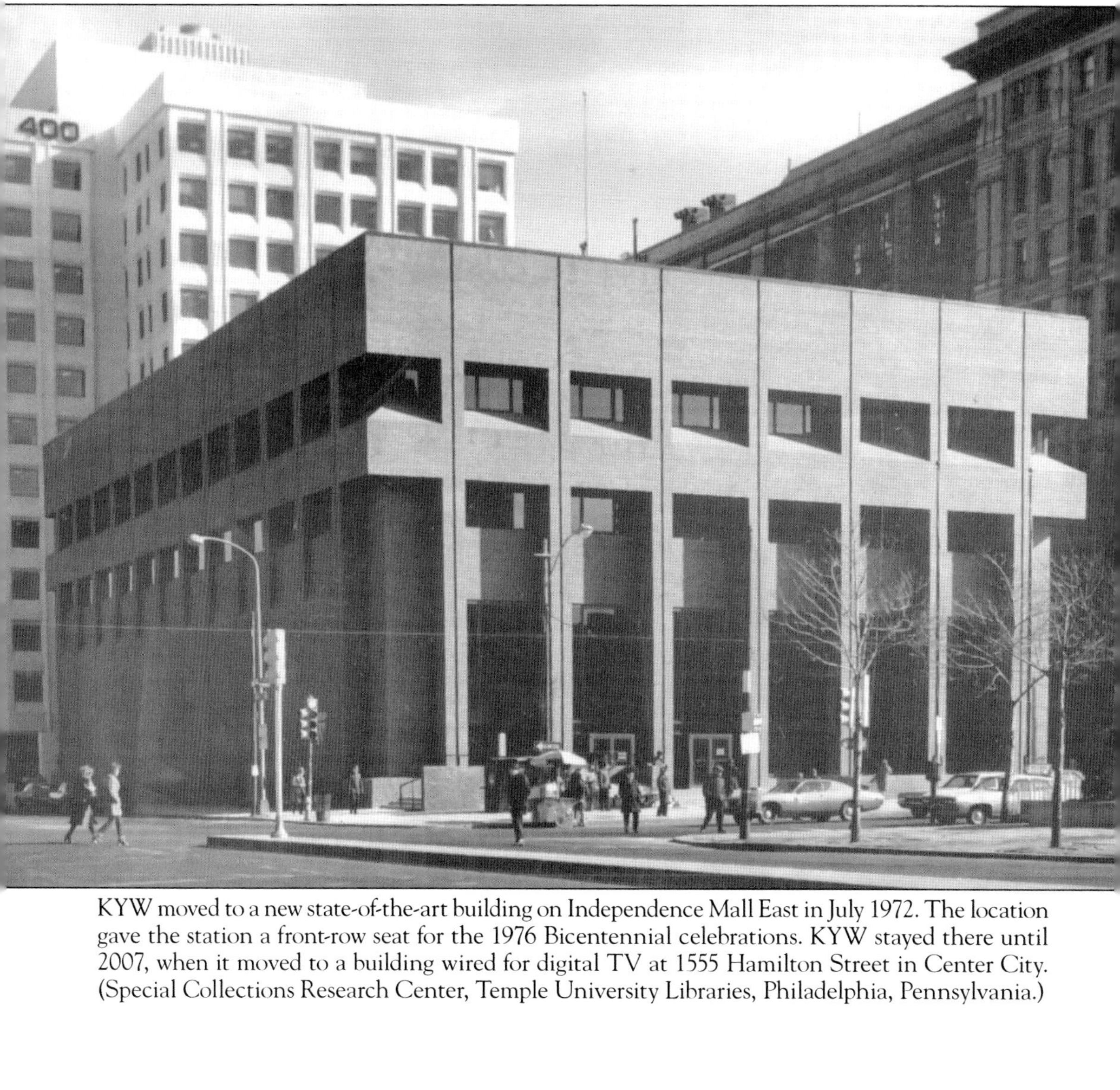

KYW moved to a new state-of-the-art building on Independence Mall East in July 1972. The location gave the station a front-row seat for the 1976 Bicentennial celebrations. KYW stayed there until 2007, when it moved to a building wired for digital TV at 1555 Hamilton Street in Center City. (Special Collections Research Center, Temple University Libraries, Philadelphia, Pennsylvania.)

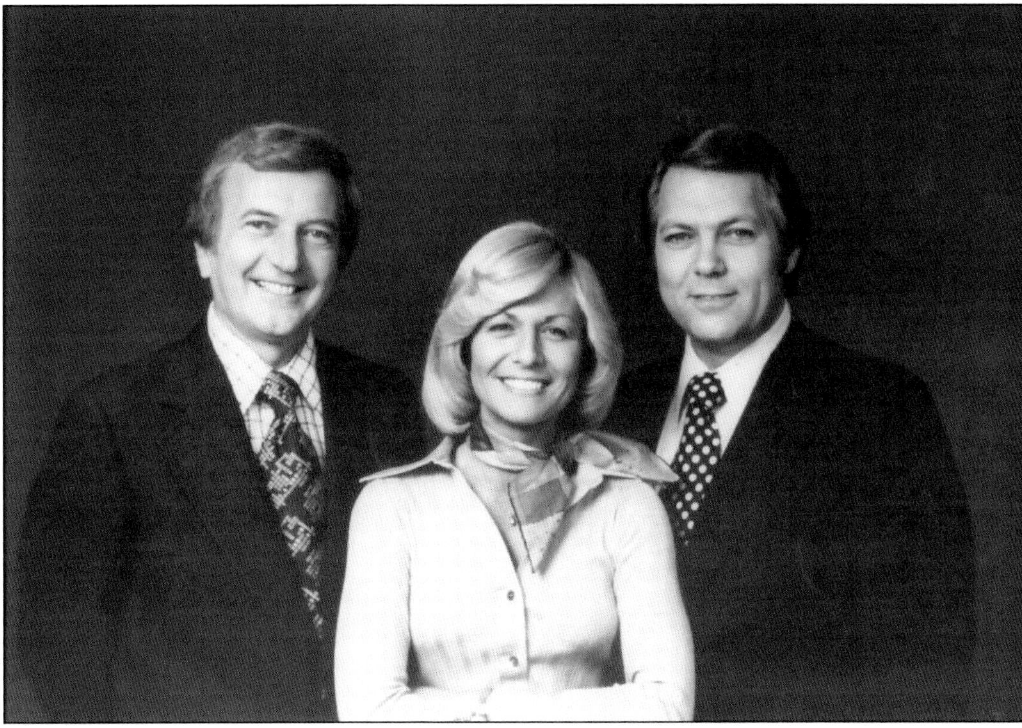

The glory days of *Eyewitness News* came in the mid-1970s when Mort Crim (right) anchored *Newswatch 5:30* with Jessica Savitch (center), and Vince Leonard (left) anchored *Eyewitness News* at 6:00 p.m. and 11:00 p.m. (KYW-TV.)

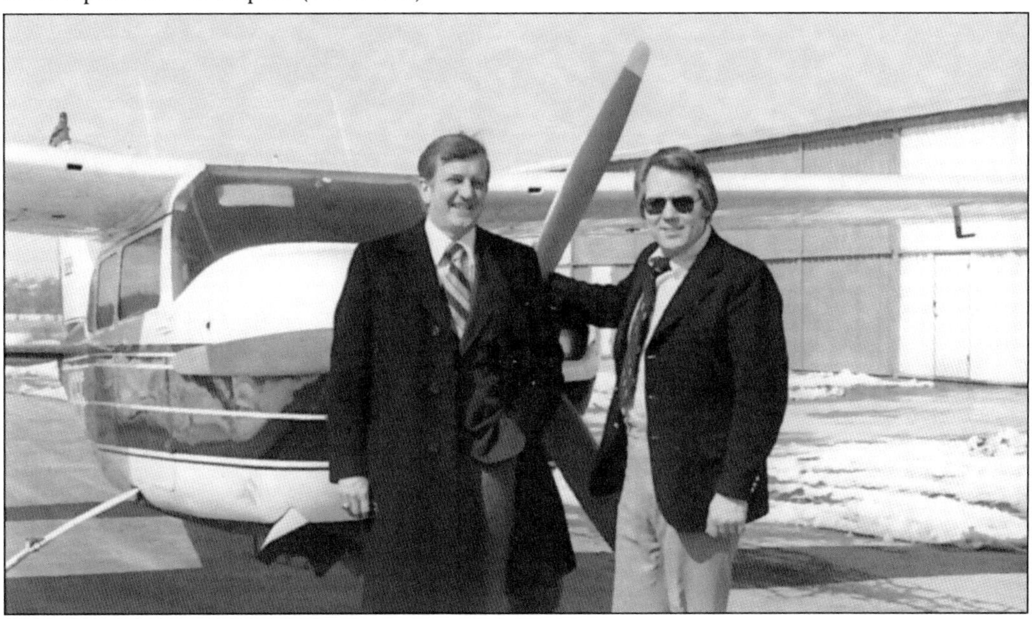

KYW anchors Vince Leonard (left) and Mort Crim (right) not only worked together, they flew together. They co-owned a Cessna that was based at Wings Field just outside of Philadelphia. Some former KYW staffers say the station did not want its two top anchors flying in the same plane at the same time. (Mort Crim.)

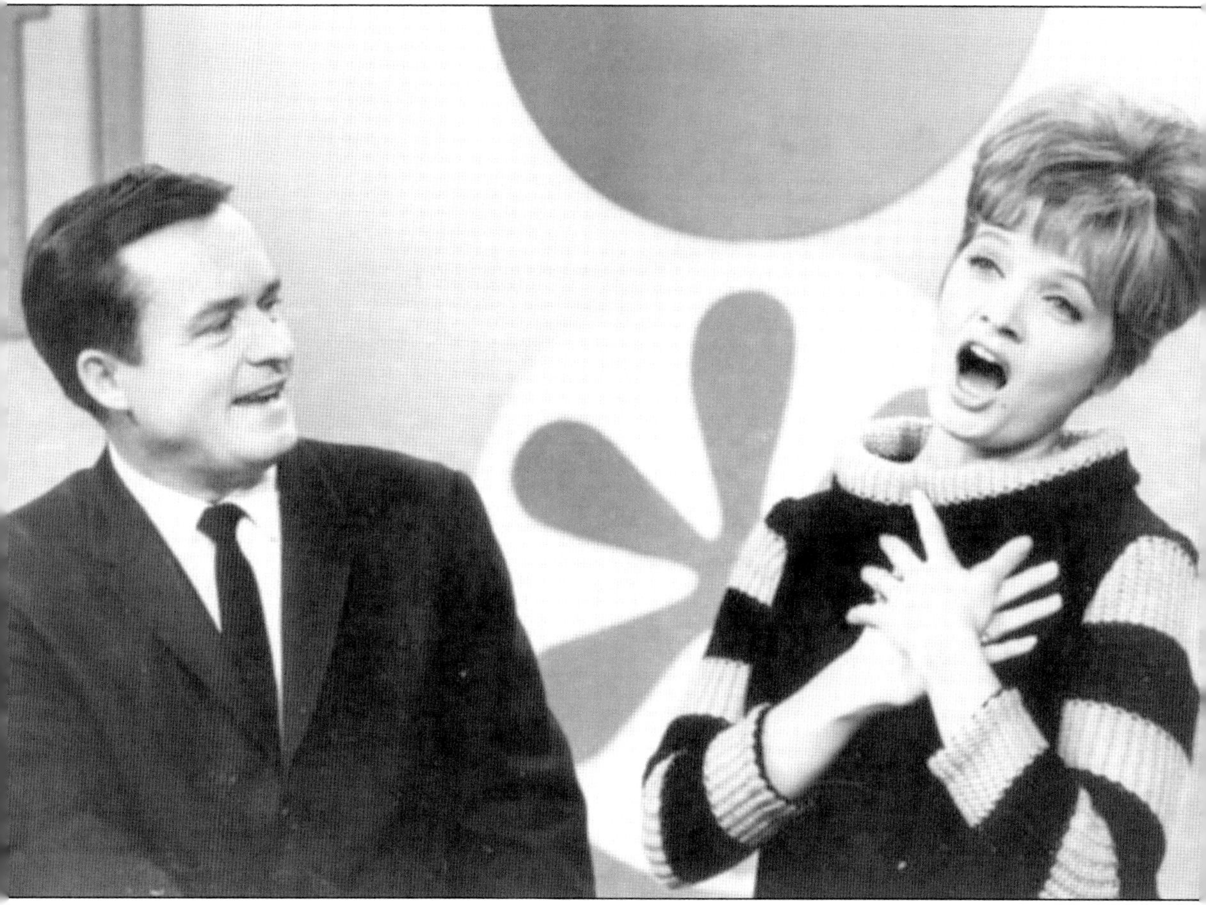

The Mike Douglas Show was the top rated afternoon talk show when it started in Cleveland. When Westinghouse bought KYW, it moved the show to Philadelphia. This picture shows Douglas singing with actress Florence Henderson. In his book *We'll Be Right Back*, Douglas said his trademark asterisk logo was "just a cute symbol that gave the set some pop." (Courtesy of the Philadelphia History Museum at the Atwater Kent, the Historical Society of Philadelphia Collection.)

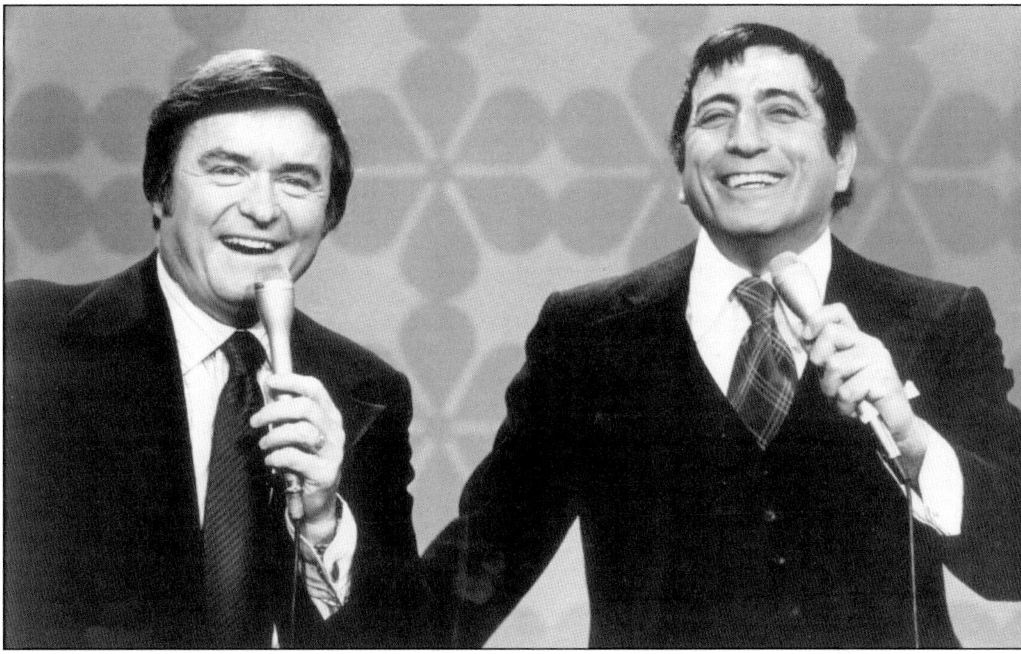

Douglas welcomed the top names in entertainment to Philadelphia, including crooner Tony Bennett. Douglas often took advantage of the station's location on Independence Mall to do segments outside and even at the nearby Liberty Bell. (Courtesy of the Philadelphia History Museum at the Atwater Kent, the Historical Society of Philadelphia Collection.)

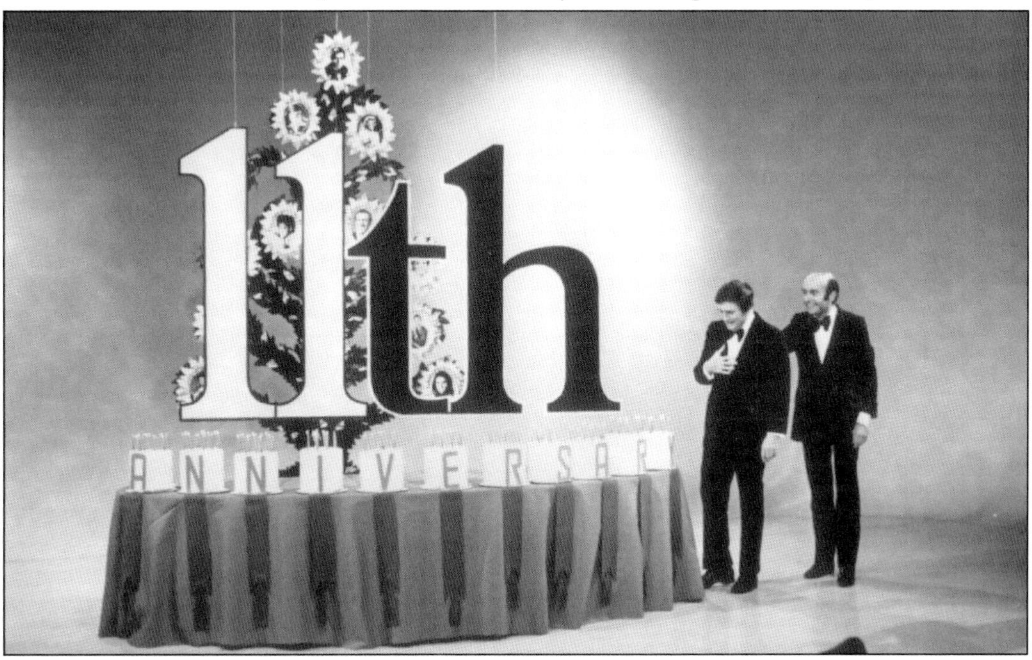

Douglas hosted his show from KYW's Walnut Street and Fifth and Market Streets studios from 1965 until he moved the show to the West Coast in 1984. The show celebrated its 11th anniversary in 1976. (Courtesy of the Philadelphia History Museum at the Atwater Kent, the Historical Society of Philadelphia Collection.)

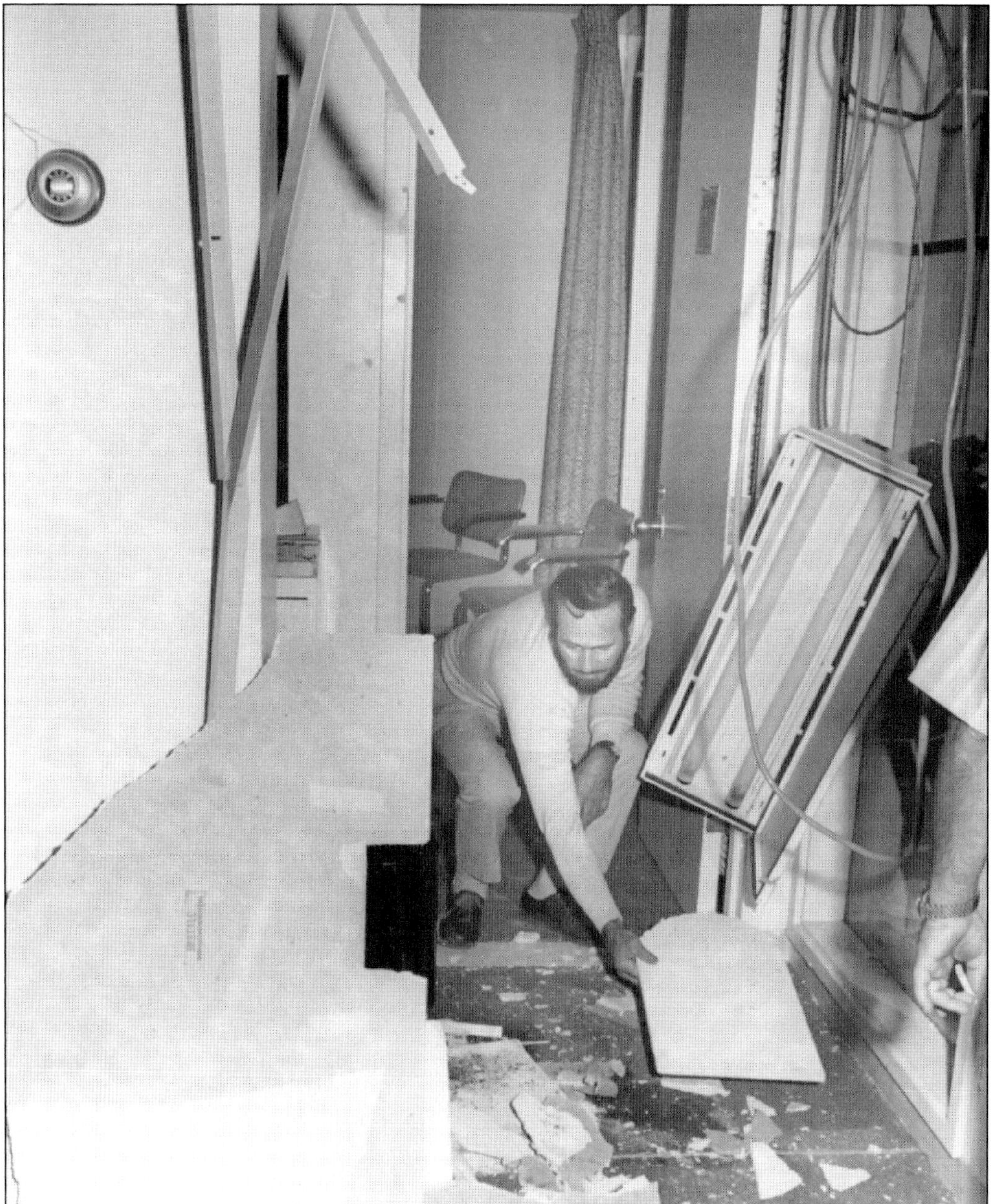

KYW's Lou Wagner cleans up damage backstage caused by a chimpanzee that went out of control on *The Mike Douglas Show.* In his book *We'll Be Right Back*, Douglas said the chimp caused $20,000 worth of damage. Wagner later became the weekend weatherman at WCAU, known as "Captain Lou." (Special Collections Research Center, Temple University Libraries, Philadelphia, Pennsylvania.)

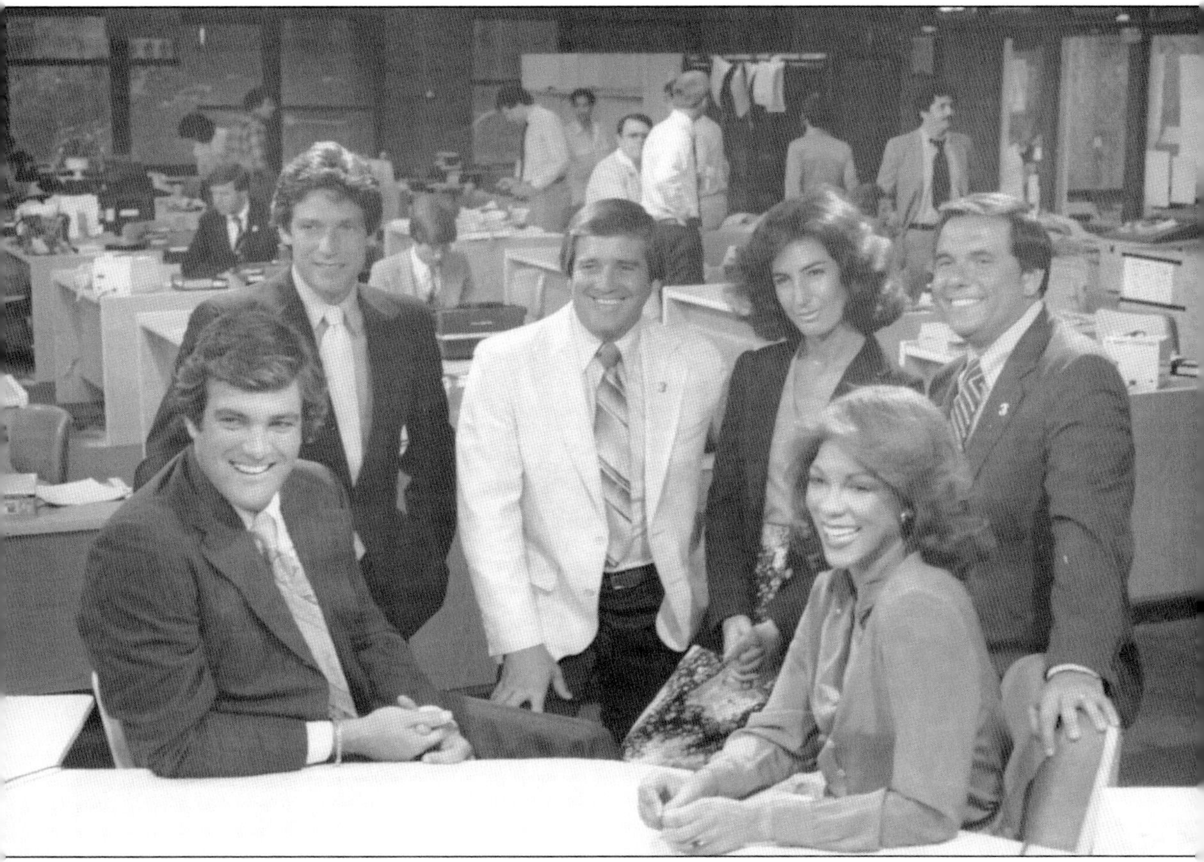

KYW's news team in 1980 was made up of, standing from left to right, Maury Povich (host of *People Are Talking* and a news anchor), sports director Bob Dominie, meteorologist/science reporter Leah Erickson, and anchor Dick Sheeran. Seated at the desk are anchors Patrick Emory and Beverly Williams, who would later anchor together at CNN. (KYW)

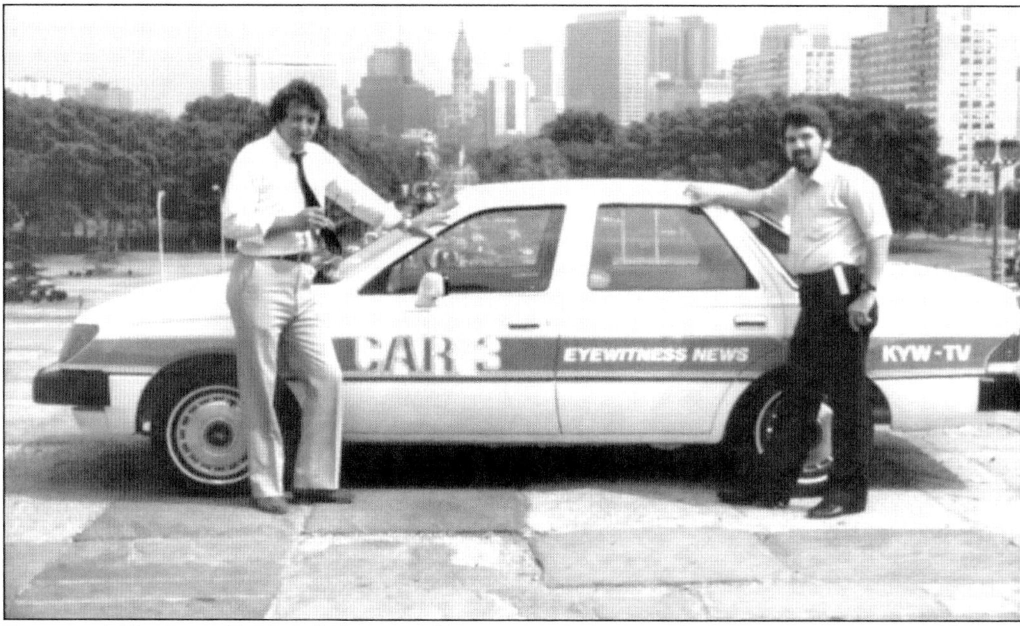

Reporter Mike Strug (left) and photographer Irv Grozky pose in front of "Car 3." The duo traveled around the Delaware Valley looking for interesting stories. Car 3 was unveiled with a lot of fanfare outside of KYW's studios during *Eyewitness News at Noon* in 1983. (KYW-TV.)

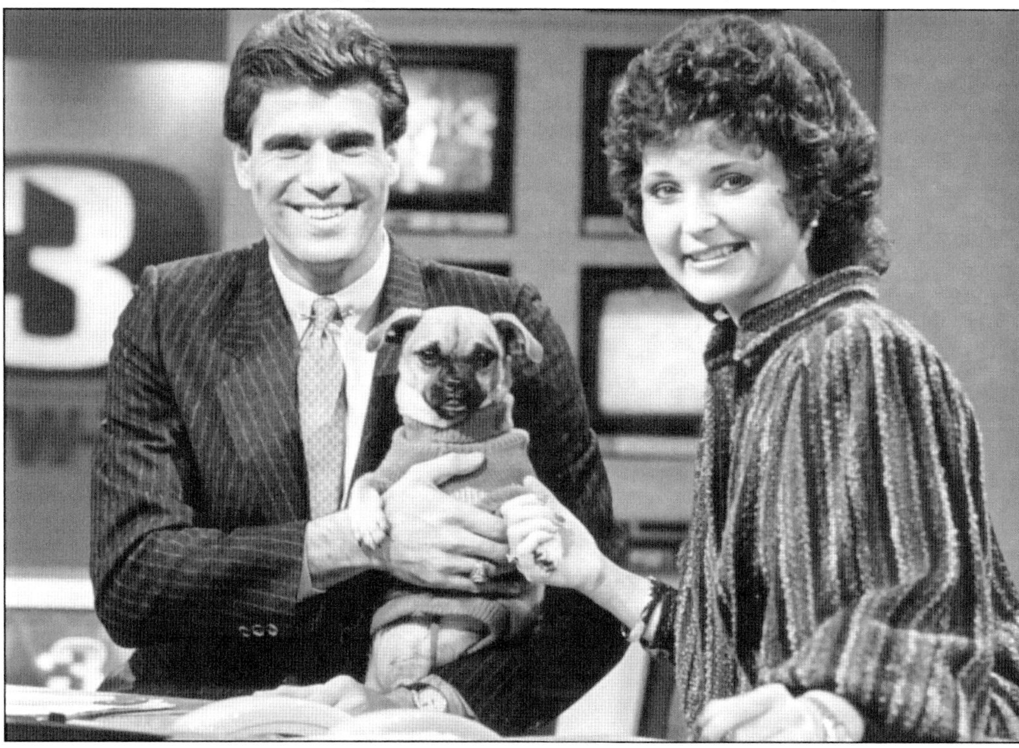

Jerry Penacoli and Pat Ciarrocchi anchored KYW's *Eyewitness News at Noon* in the 1980s. They hosted a pet adoption segment every Friday. Penacoli went on to host several nationally syndicated entertainment shows. Ciarrocchi is Philadelphia's longest-tenured female anchor. (KYW-TV.)

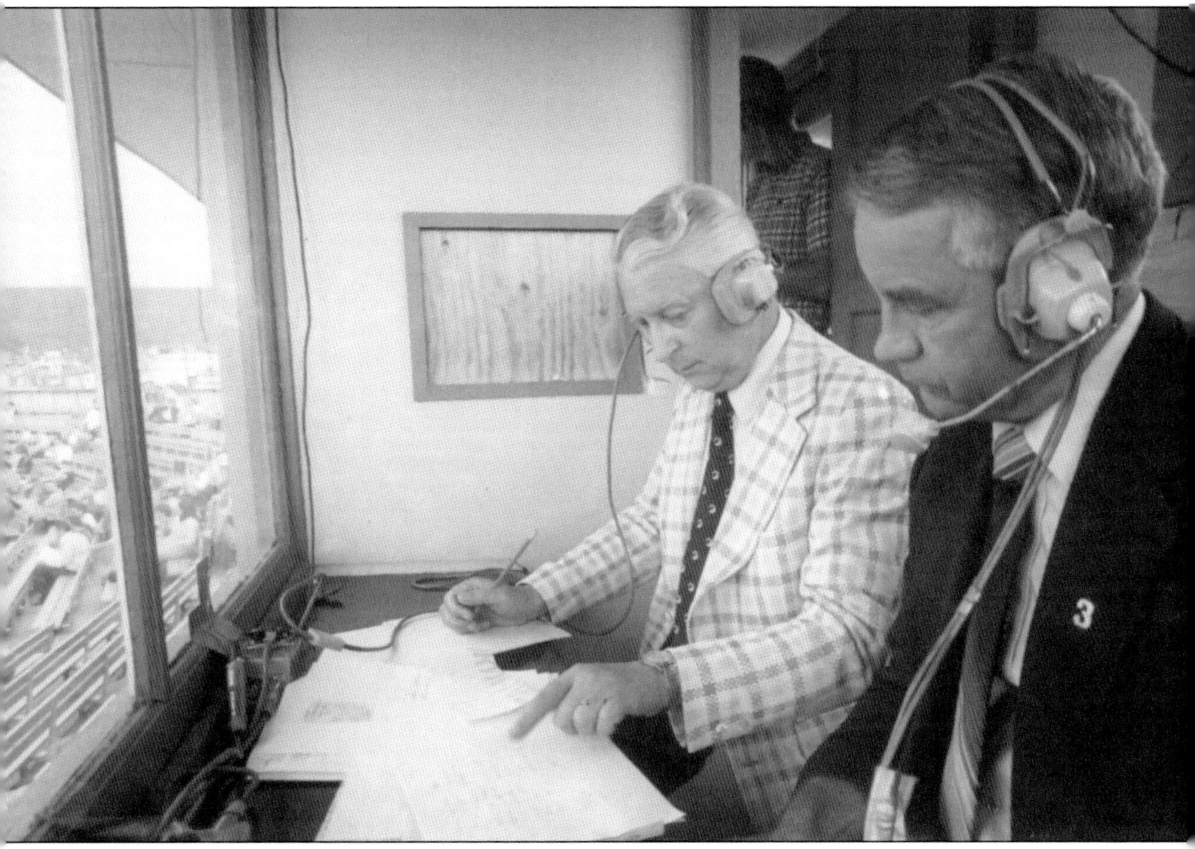

Bob Bradley (left) worked at WRCV/KYW from 1956 to 1987. He was primarily known as a sportscaster, but began as the children's show host Buckskin Billy, playing the titular character. Bradley did play-by-play for Phillies spring training games for many years. This 1981 photograph shows him in the broadcast booth at Jack Russell Stadium in Clearwater, Florida, along with former Phillies coach Billy DeMars, who is wearing a KYW lapel pin that was also worn by the news anchors at the time. (Special Collections Research Center, Temple University Libraries, Philadelphia, Pennsylvania.)

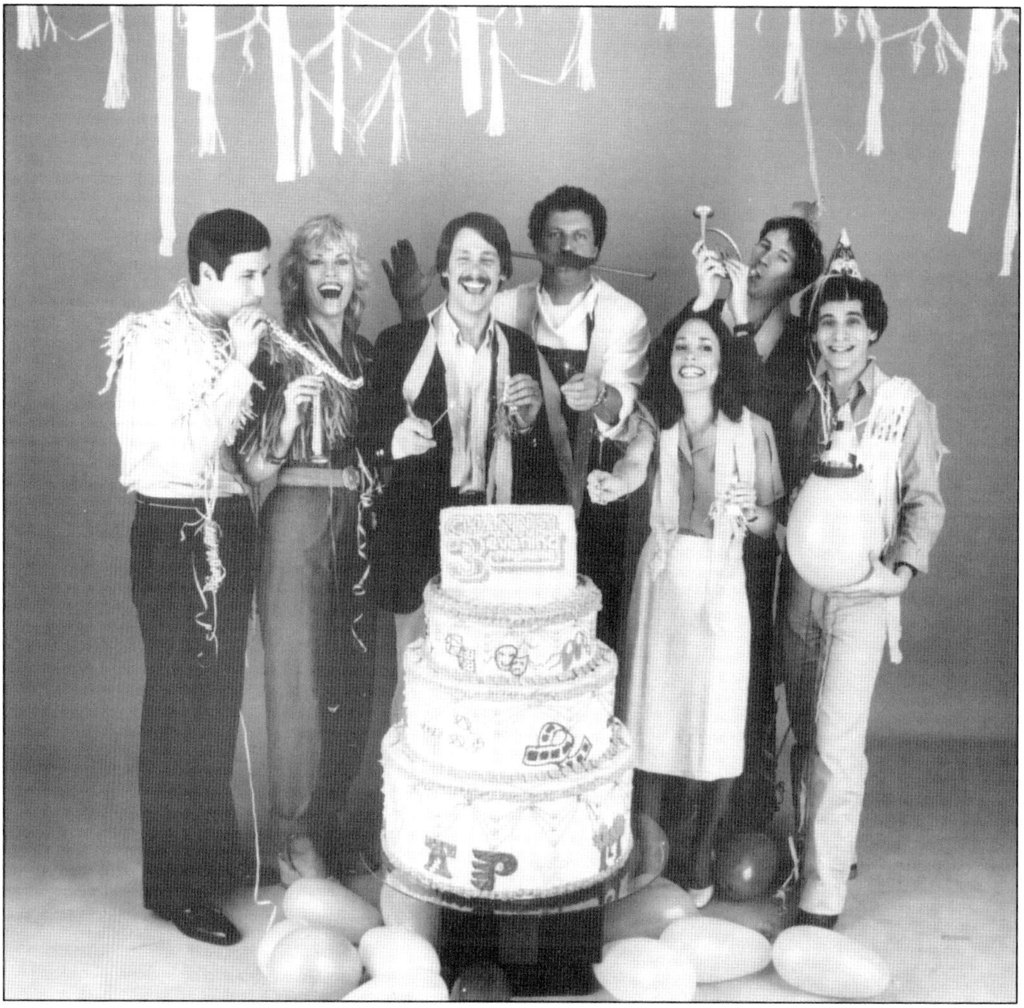

Evening Magazine aired at 7:30 nightly on KYW. The show celebrated its third birthday in 1990 with, from left to right, Dr. Marc Rosenberg (medical contributor), Judy Barton (movie reviewer), cohost Ray Murray, Chef Tell Erhardt (cooking contributor), cohost Suzie Pevaroff, and "Discovery Tipsters" Bob Meyer and Bob Young. The show was canceled in 1992. *Evening Magazine* was a franchise that aired on stations across the country. (Special Collections Research Center, Temple University Libraries, Philadelphia, Pennsylvania.)

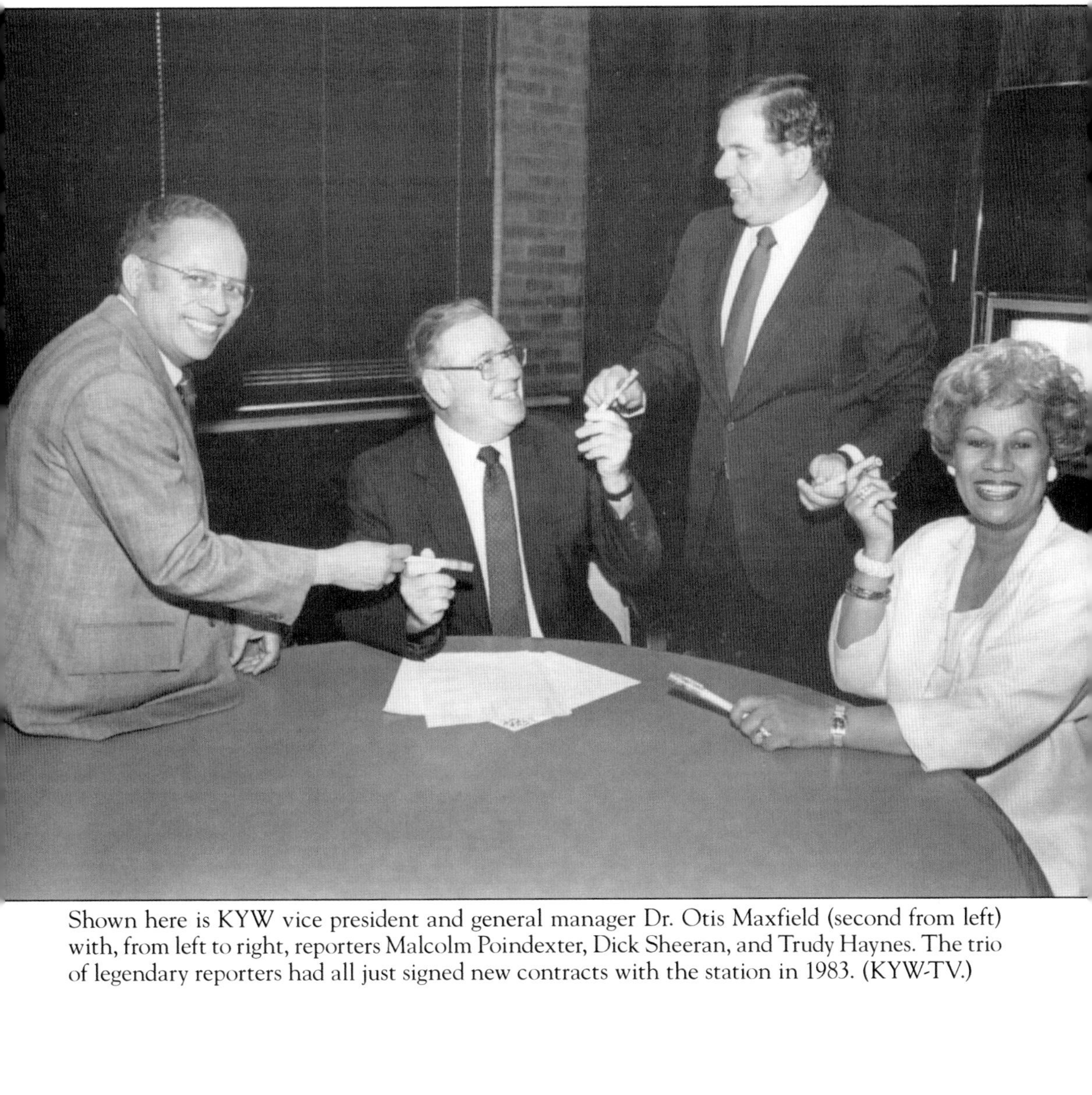

Shown here is KYW vice president and general manager Dr. Otis Maxfield (second from left) with, from left to right, reporters Malcolm Poindexter, Dick Sheeran, and Trudy Haynes. The trio of legendary reporters had all just signed new contracts with the station in 1983. (KYW-TV.)

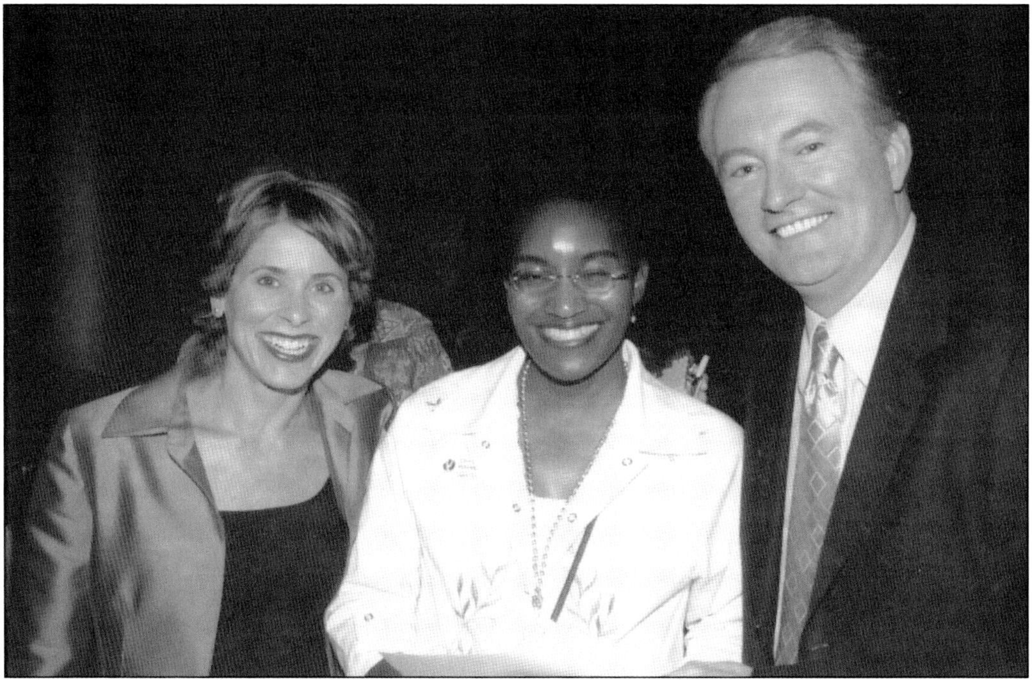

KYW meteorologist Kathy Orr (left) and anchorman Larry Mendte flank veteran WCAU producer Catherine Brown at a local Emmy Awards event. Both Orr and Mendte also worked at NBC10 before heading to Channel 3 in the early 2000s. (National Academy of Television Arts and Sciences.)

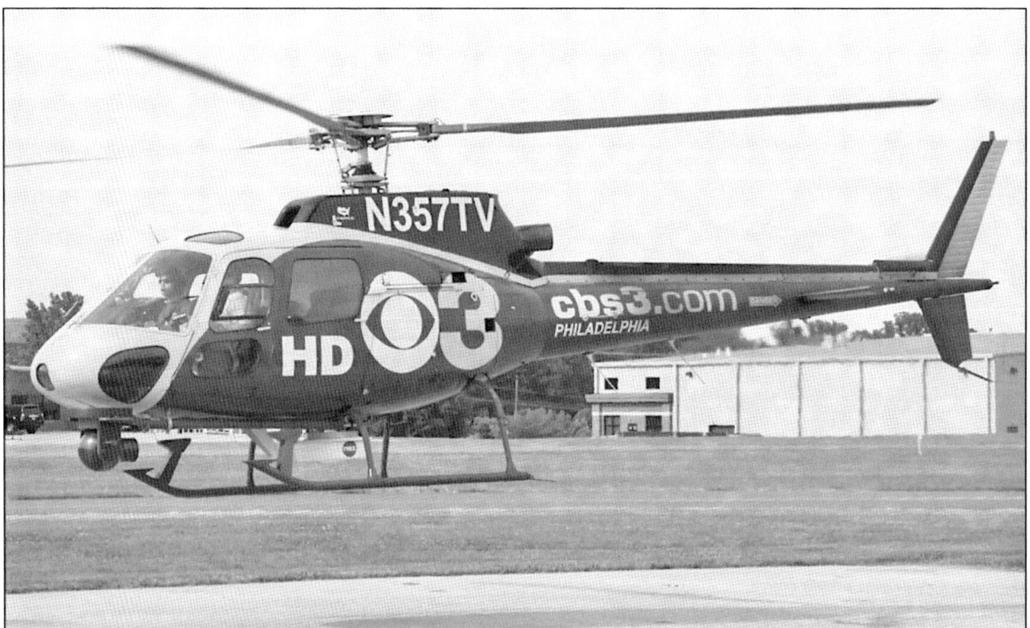

KYW's Chopper 3 takes off for another breaking news story. Channel 3 shares its helicopter with WTXF-TV as part of a video sharing arrangement called Local News Service (LNS). (Author's collection.)

Two

WFIL/WPVI Channel 6

WFIL-TV's first studios were in the Widener Building at Broad and Chestnut Streets in Center City. The station was owned by the *Philadelphia Inquirer.* In 1947, Channel 6 moved to a new, state-of-the-art building at Forty-sixth and Market Streets in West Philadelphia. It was the first building in the world built exclusively to be a television broadcasting facility, according to WFIL. The station again moved in 1964, this time to a round building at 4100 City Avenue designed by architect Vincent Kling. The station was sold to Capital Cities Broadcasting in 1969. The call letters were changed to WPVI in 1971. The most well known personality to start at Channel 6 is by far *Bandstand* host Dick Clark. He took over the teenage dance party show from Bob Horn and Lee Stewart in 1959. Clark moved the show to Los Angeles in 1964, changed the name to *American Bandstand*, and built an empire. Channel 6 is known for its local legends too. Viewers from several generations cannot think of Channel 6 without thinking of *Captain Noah & His Magical Ark*, *The Chief Halftown Show*, *The Larry Ferrari Show*, *Sally Starr*, and *Al Alberts' Showcase*. Channel 6 *Action News* has been the dominant newscast in Philadelphia for more than 45 years. The format started in April 1970 with anchorman Larry Kane and a catchy theme song called "Move Closer To Your World." The team of Kane, weatherman Jim O'Brien, and sportscaster Joe Pellegrino became a TV dynasty. *Action News* has rarely dropped out of first place in any time slot since. Anchor Jim Gardner has continued the dominance since 1978, when Kane left for New York. WPVI is currently owned by Disney/ABC and is still located at 4100 City Avenue.

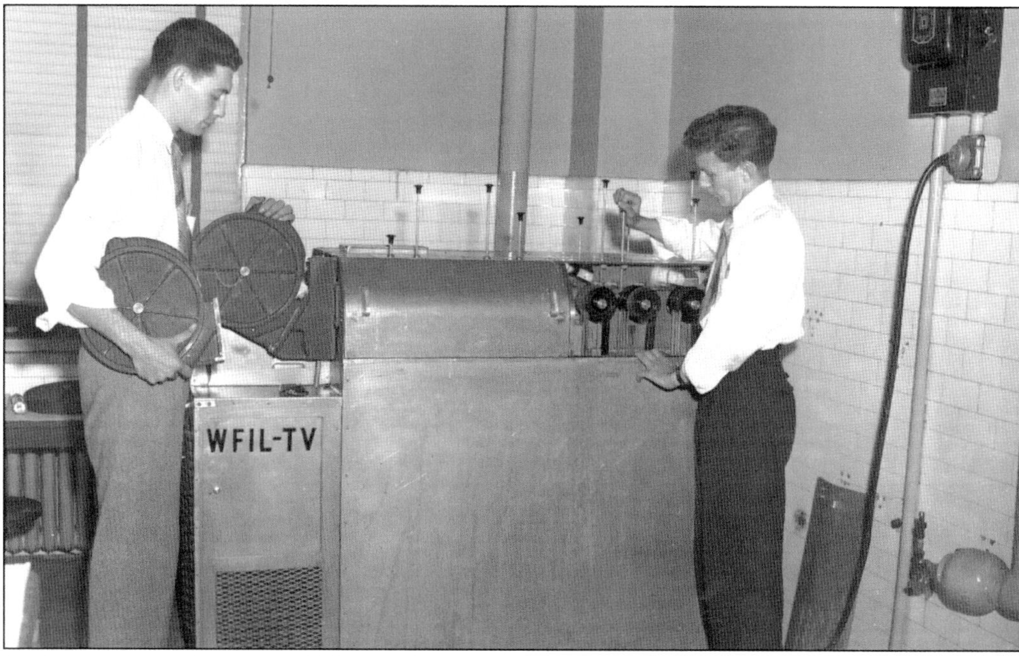

Technicians develop film at WFIL-TV's first studios in 1948. The station was originally located on the 18th, then 11th floors, of the Widener Building at Broad and Chestnut Streets in Center City. (Courtesy of Special Collections Research Center, Temple University Libraries, Philadelphia, Pennsylvania.)

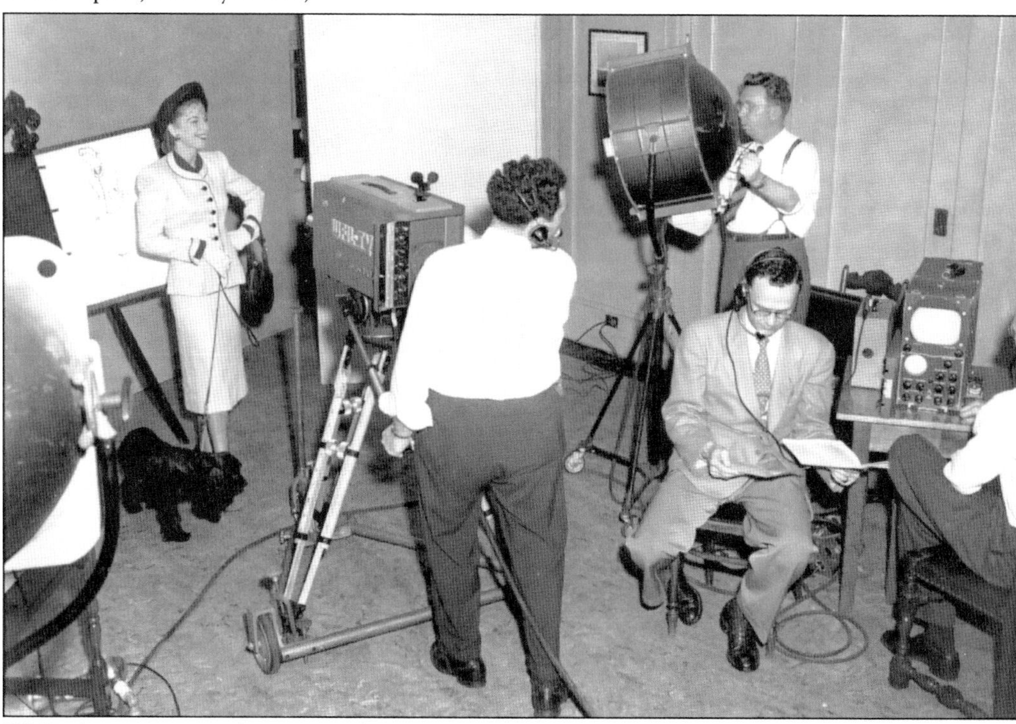

WFIL demonstrates television at the University Club in Center City, May 23, 1947. (Courtesy of Special Collections Research Center, Temple University Libraries, Philadelphia, Pennsylvania.)

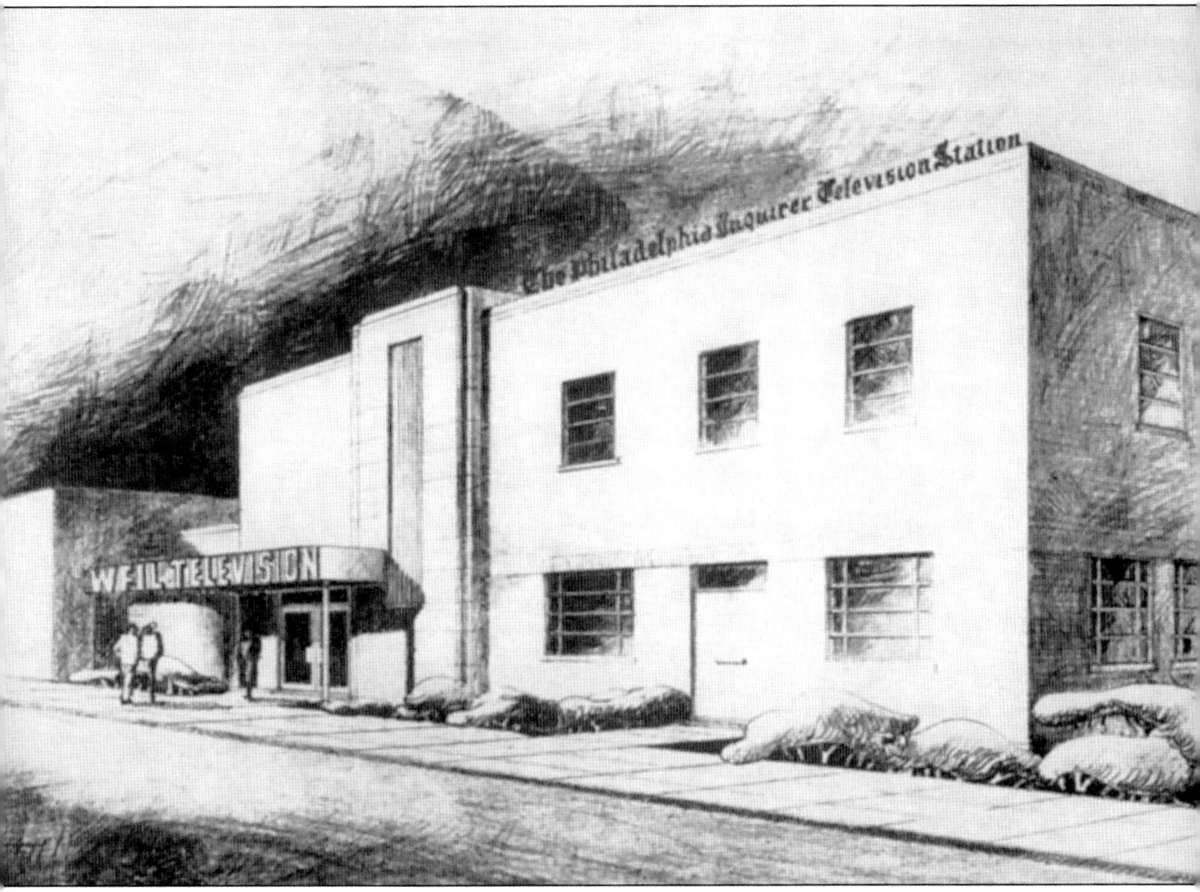

This is an early sketch of WFIL-TV's building at Forty-sixth and Market Streets in West Philadelphia, which opened in 1947. The station was owned by the *Philadelphia Inquirer*. WFIL has always claimed the building was the first in the world to be designed to be a TV station. (WPVI-TV.)

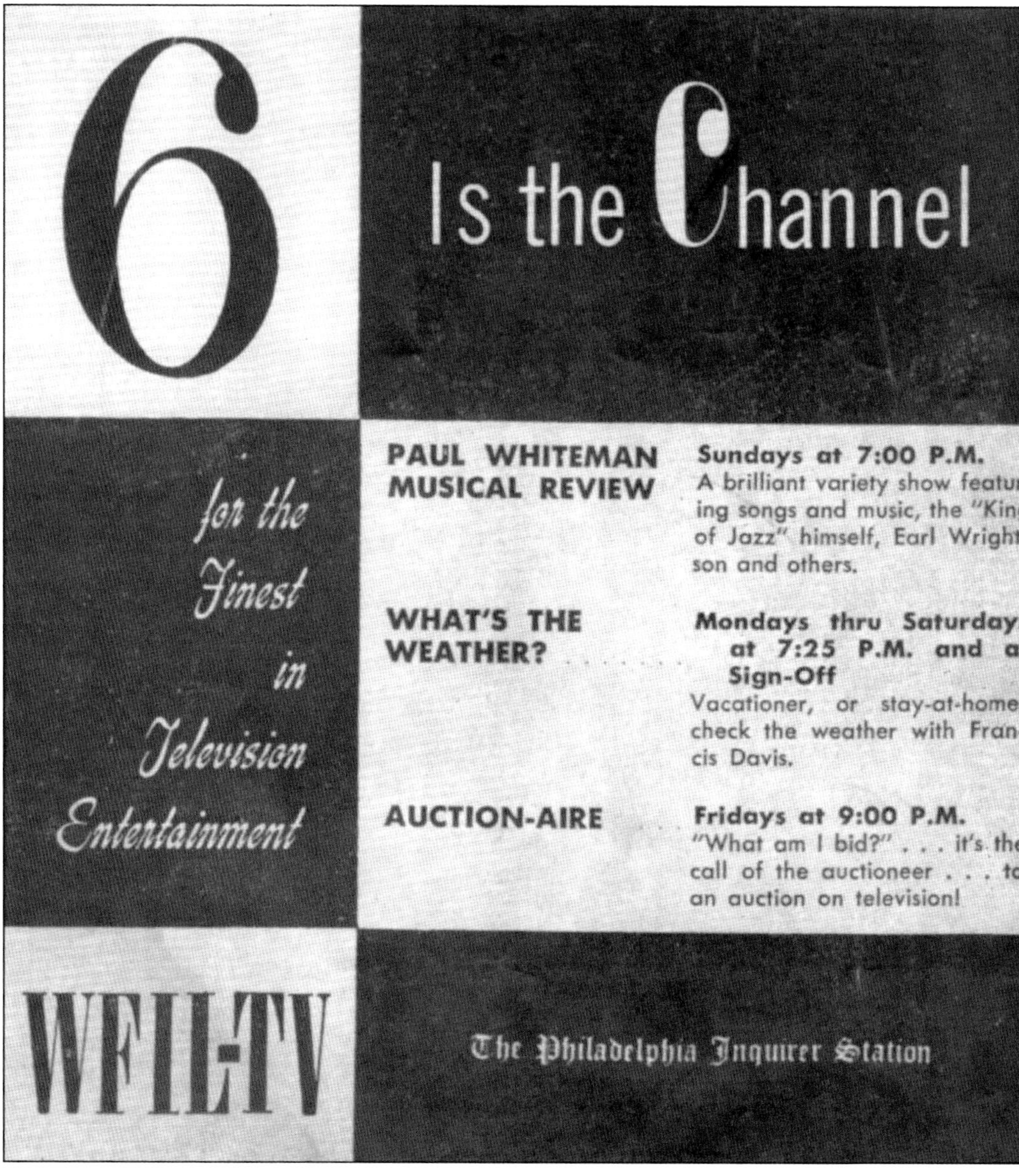

This is a WFIL-TV Channel 6 advertisement promoting three programs for the week of June 3, 1959. The weather program was hosted by Francis Davis, Philadelphia's first on-air meteorologist. (WPVI-TV.)

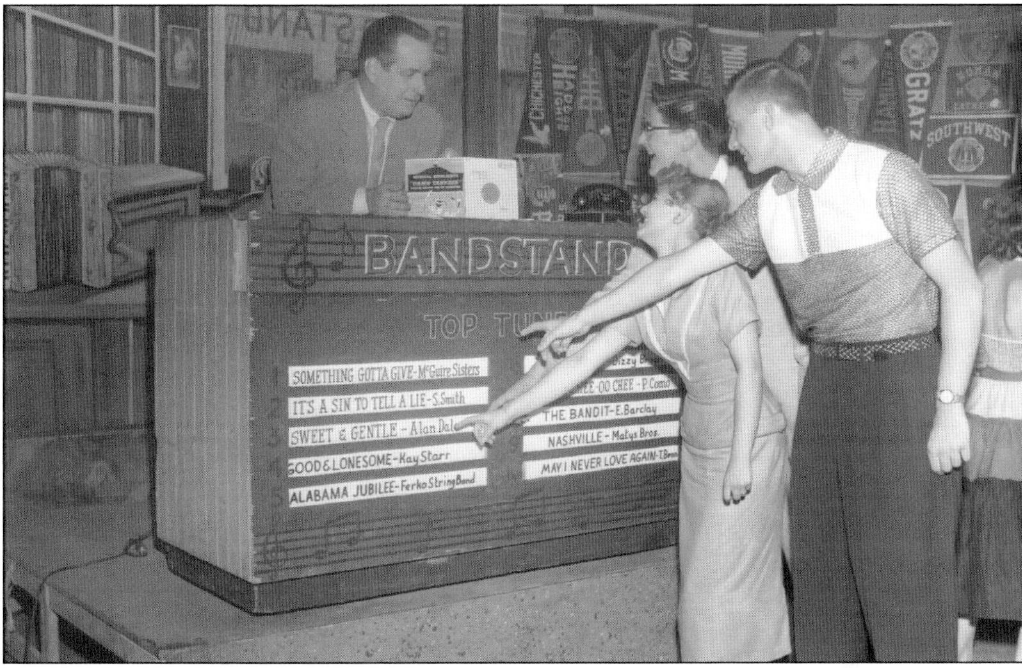

Teenagers gather around the desk of *Bandstand* host Bob Horn at WFIL's Forty-sixth and Market Streets studios on July 17, 1955. Horn and cohost Lee Stewart were the first hosts of the iconic dance show. In 1956, Horn was fired after a drunk-driving arrest, and Lee was let go. They were replaced with a young disc jockey from Syracuse named Dick Clark, and the rest is history. (Special Collections Research Center, Temple University Libraries, Philadelphia, Pennsylvania.)

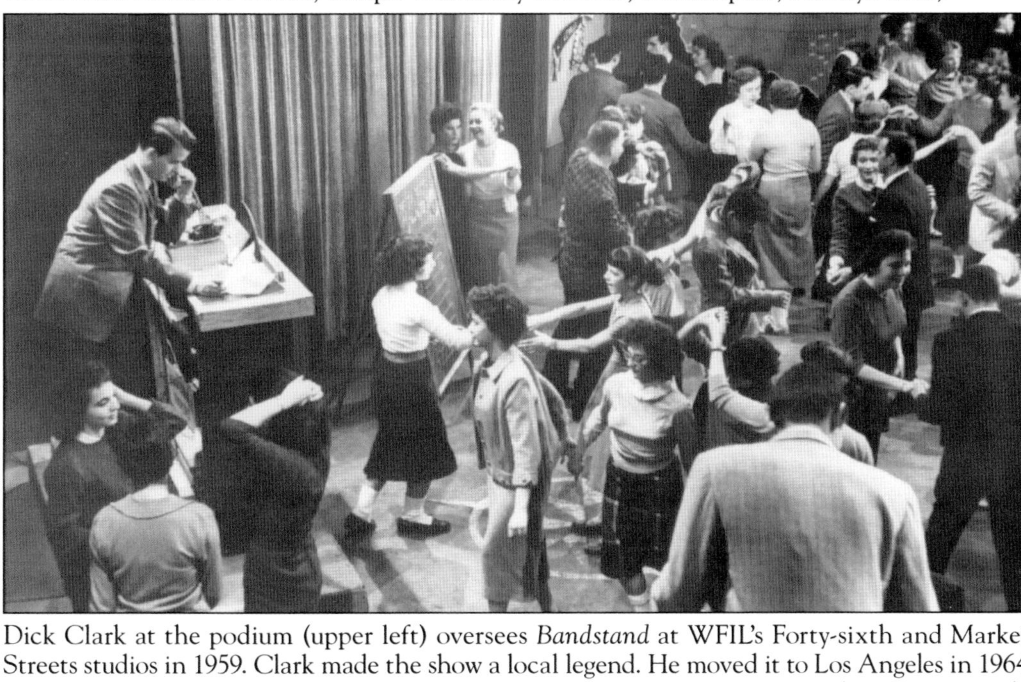

Dick Clark at the podium (upper left) oversees *Bandstand* at WFIL's Forty-sixth and Market Streets studios in 1959. Clark made the show a local legend. He moved it to Los Angeles in 1964 and changed the name to *American Bandstand*. (Special Collections Research Center, Temple University Libraries, Philadelphia, Pennsylvania.)

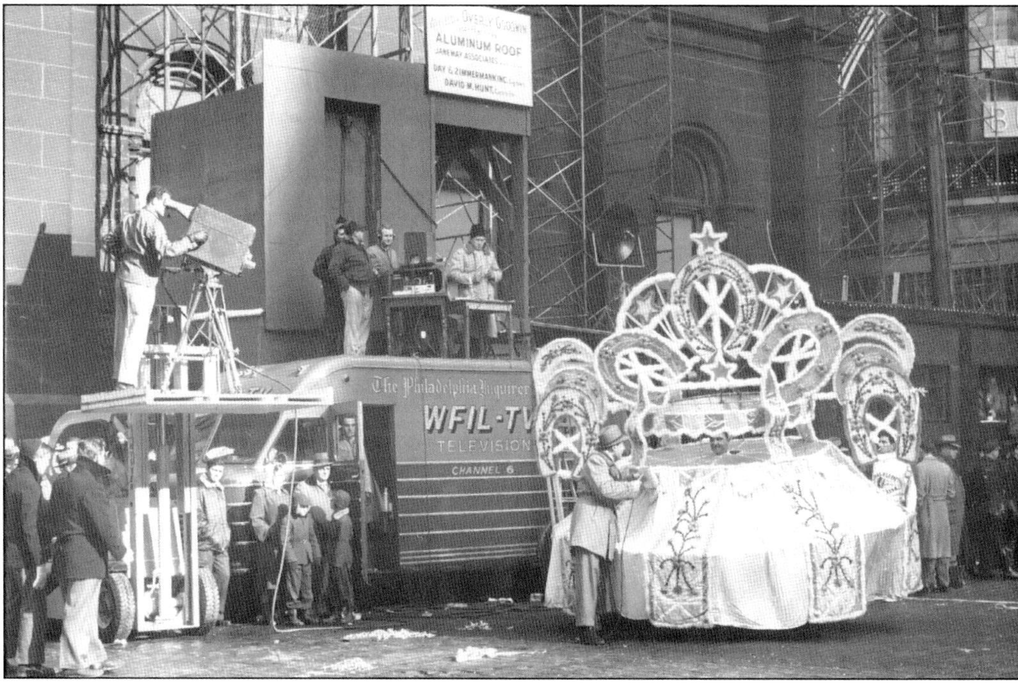

A photograph from the mid-1950s shows WFIL's broadcast location for the annual New Year's Day Mummers Parade. Channel 6 has also broadcast the Thanksgiving Day parade and many other parades over the decades. (Special Collections Research Center, Temple University Libraries, Philadelphia, Pennsylvania.)

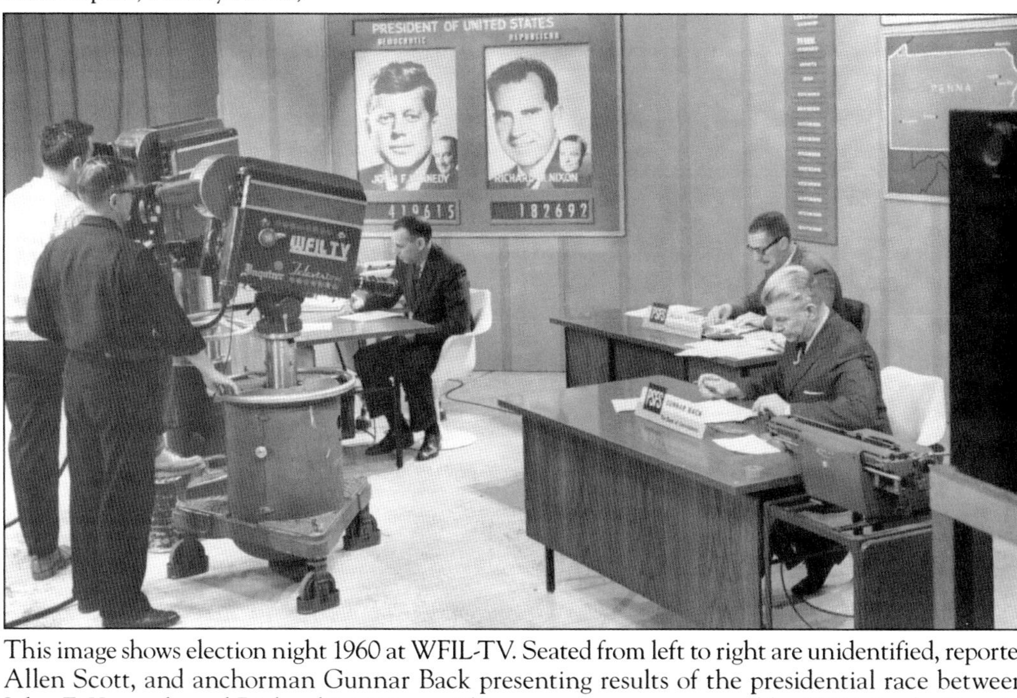

This image shows election night 1960 at WFIL-TV. Seated from left to right are unidentified, reporter Allen Scott, and anchorman Gunnar Back presenting results of the presidential race between John F. Kennedy and Richard M. Nixon. The coverage was sponsored by PSFS Bank. (Special Collections Research Center Temple University Archives Philadelphia, Pennsylvania.)

Rex Morgan was a popular host on WFIL-TV between 1961 and 1969. His most popular show was *Morgan in the Morning*, which aired on weekdays. It was set in the fictional town of Morganville. Here, Morgan exercises with his sidekick and real-life pet dog, Horatio. (Rex Morgan Jr.)

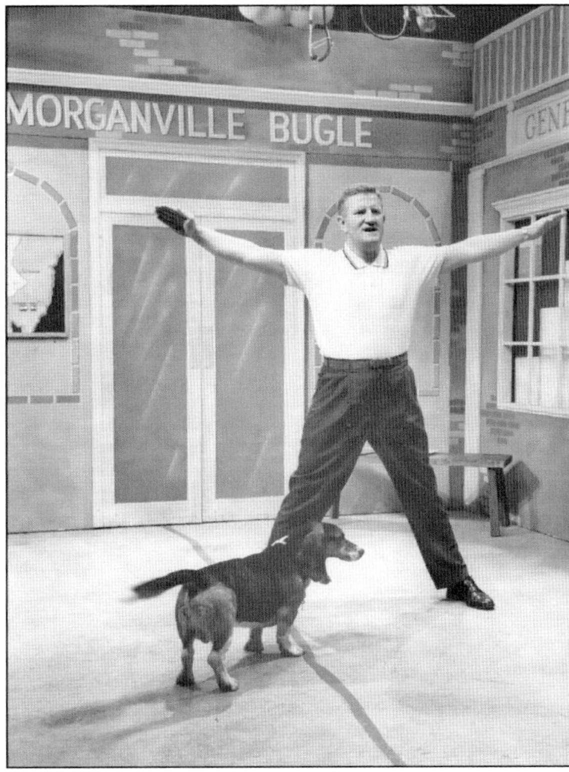

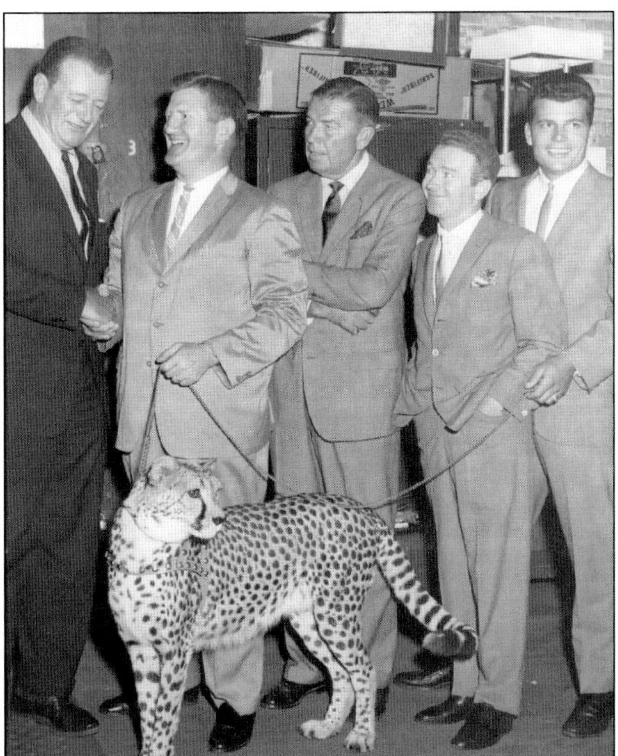

Morgan in the Morning was the go-to show for major celebrities who came to Philadelphia. Here, Morgan (second from left) welcomes, from left to right, John Wayne, Bruce Cabot, Red Buttons, and Patrick Wayne, along with a four-legged guest. (Rex Morgan Jr.)

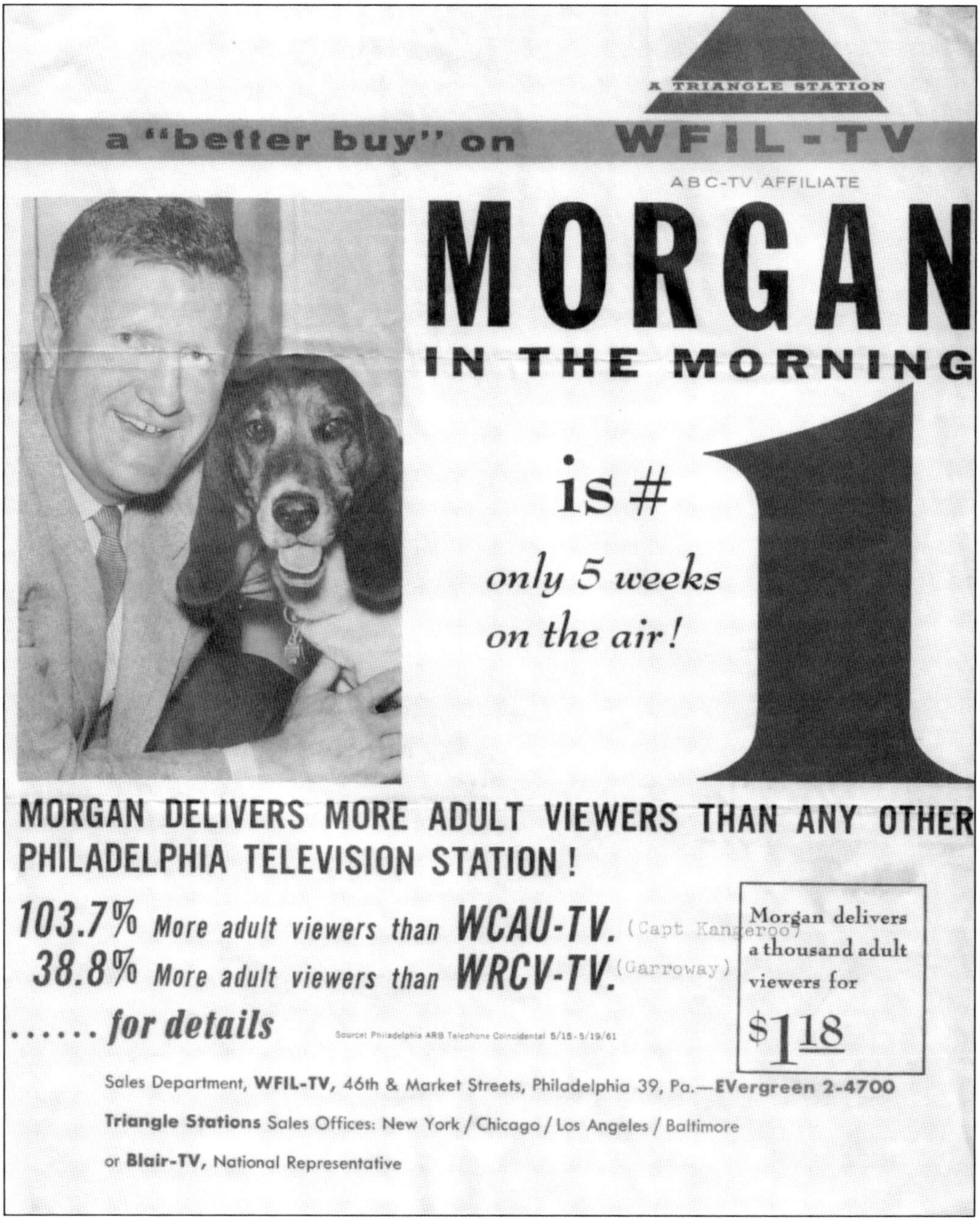

This is a rate card from 1961 showing the immediate impact of *Morgan in the Morning* on WFIL-TV. Rex Morgan also hosted a kid's show called *Ranger Rex*, which was taped at a farm in Newtown Square. In addition, Morgan filled in for fellow Channel 6 host Sally Starr when she was on vacation. (Rex Morgan Jr.)

Traynor Ora Halftown was *Chief Halftown* on Channel 6 from 1950 to 1999. When the show ended, it was the longest running children's TV show in the world. Halftown was a full-blooded Seneca from New York State. He was also a professional bowler. (WPVI-TV.)

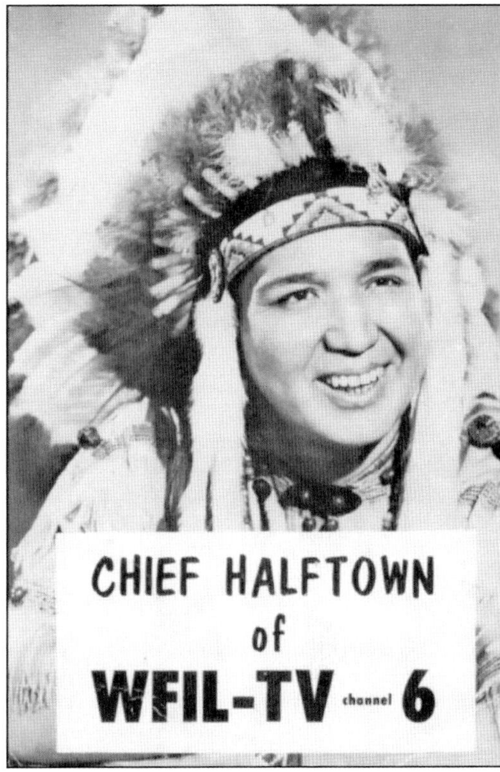

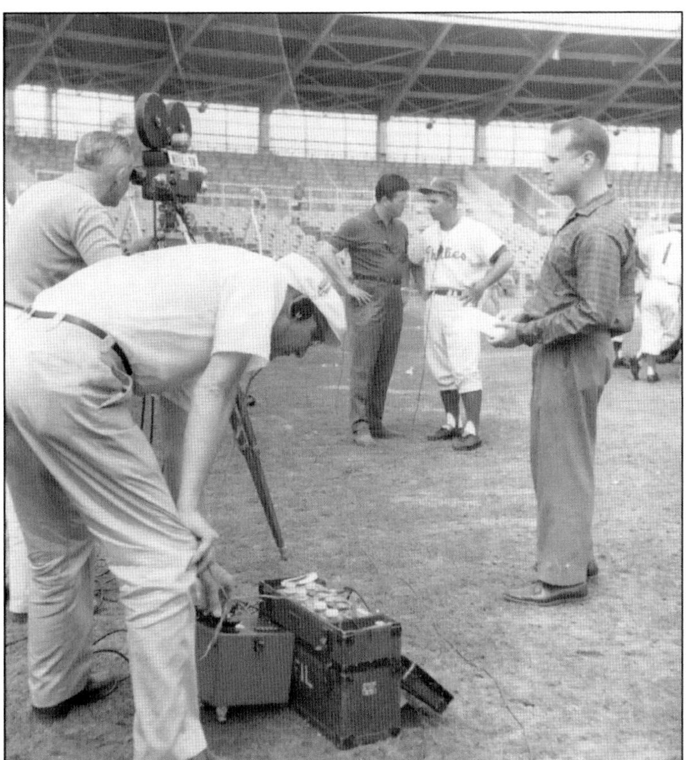

WFIL-TV was the Philadelphia Phillies' flagship station from 1959 to 1971. Program manager Lew Klein (right) oversees a Channel 6 crew covering Phillies' spring training in Clearwater, Florida, in 1963. (Special Collections Research Center, Temple University Libraries, Philadelphia, Pennsylvania.)

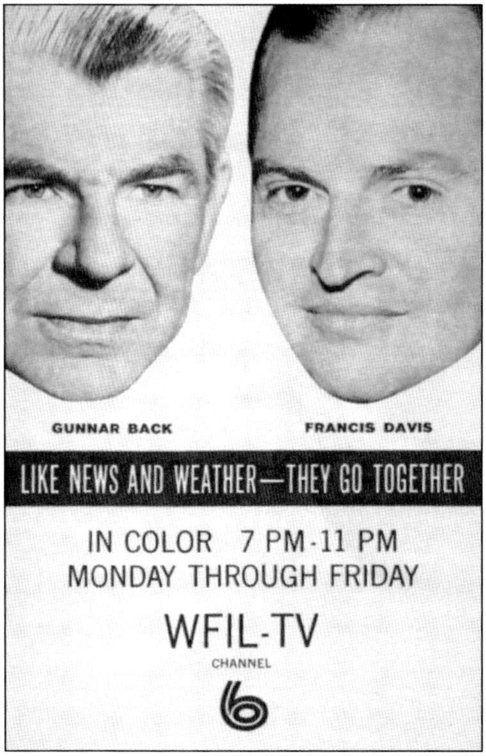

This July 1961 promo features the Channel 6 news team of anchorman Gunnar Back and meteorologist Francis Davis. Back was the station's main anchor from 1956 to 1966. He retired from Channel 6 in 1971. Davis was the station's meteorologist from 1948 to 1971. He was the first TV meteorologist to earn the Seal of Approval from the American Meteorological Society. Davis was also the dean of Drexel's College of Science. (WPVI-TV.)

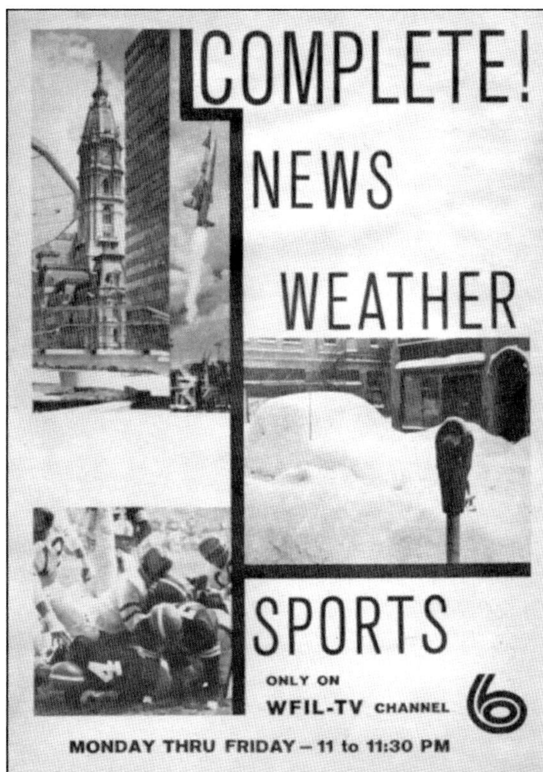

It's hard to imagine Channel 6 without *Action News*, but this promo from 1961 is a reminder of the days when WFIL was in third place behind WCAU and WRCV. (WPVI-TV.)

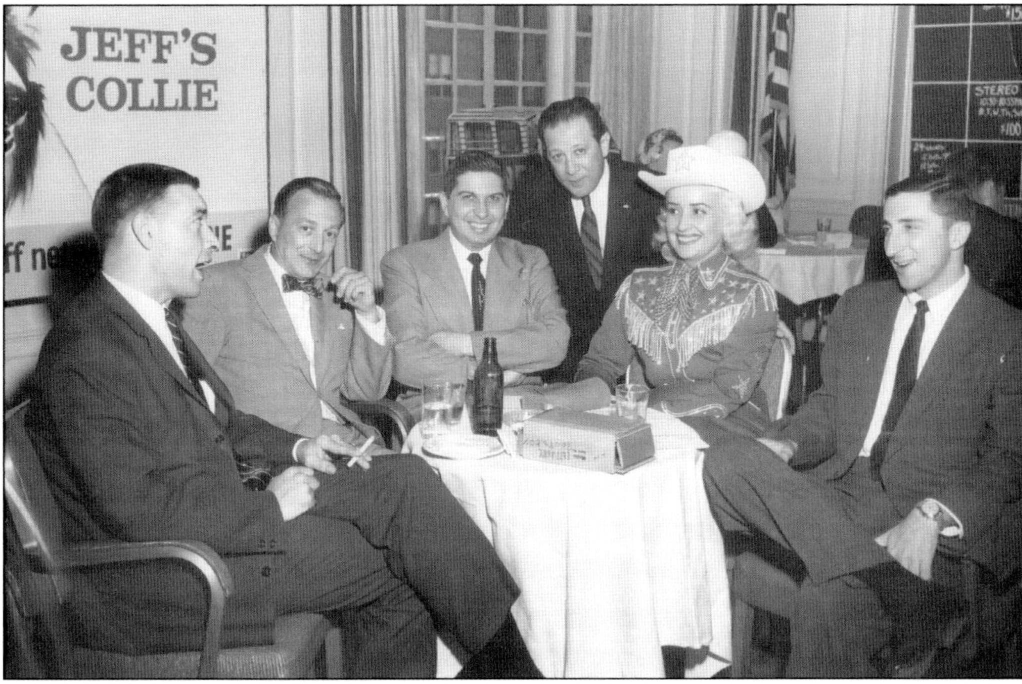

WFIL-TV salesman Max E. Solomon (standing center) and show host Sally Starr join other station employees at a party in the early 1960s. Solomon started at WFIL radio before moving over to television. (WPVI-TV.)

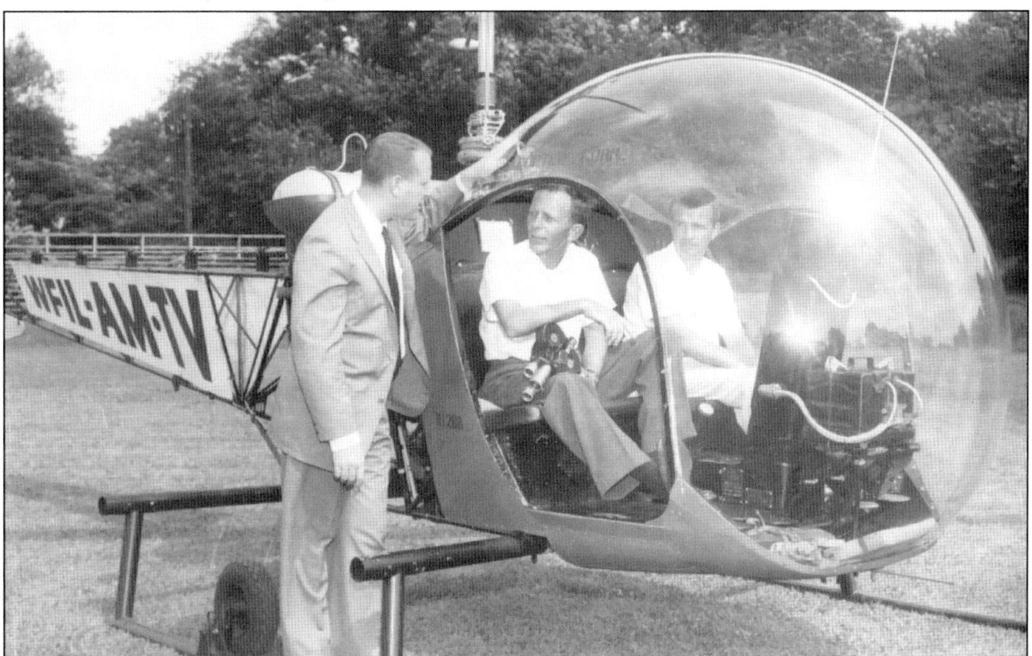

Program director Lew Klein (standing) talks with the cameraman and pilot of WFIL's news chopper in the 1960s. The Bell 47 was operated by Chester County–based Keystone Helicopter. (Special Collections Research Center, Temple University Libraries, Philadelphia, Pennsylvania.)

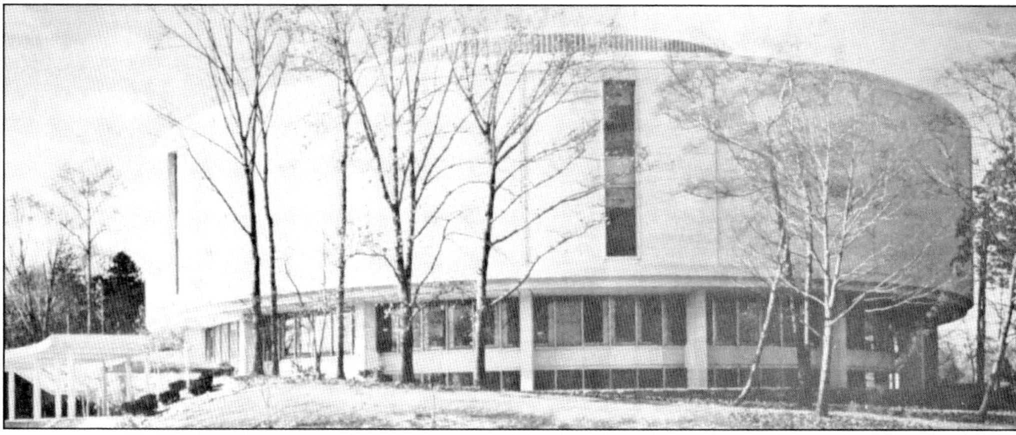

WFIL-TV moved from West Philadelphia to 4100 City Avenue in January 1964. The round building was erected for Triangle Publications, owned by Walter Annenberg. (Both, WPVI-TV.)

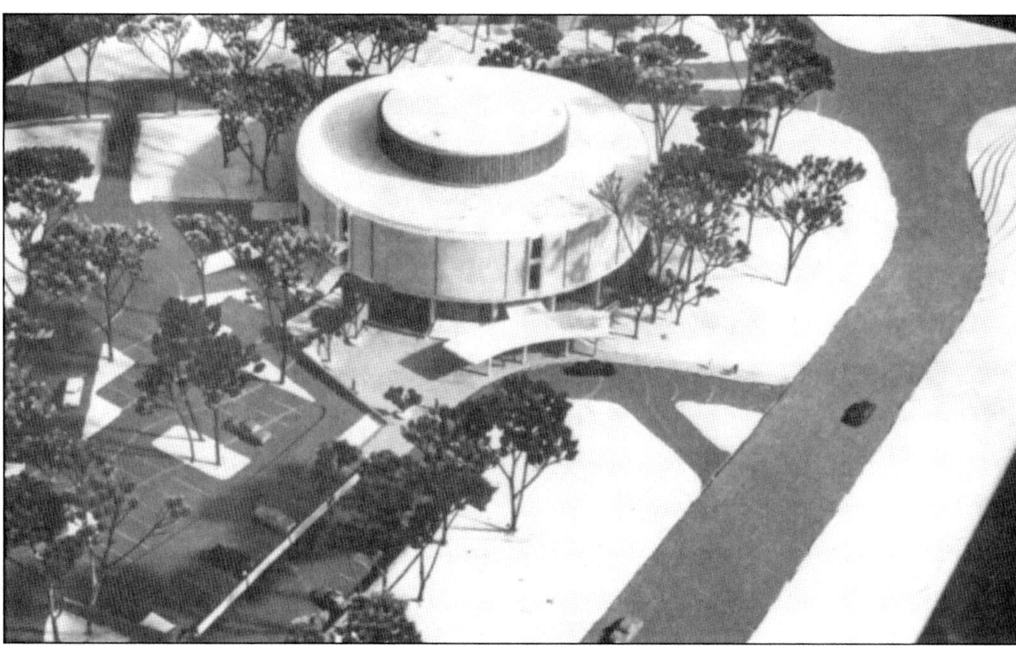

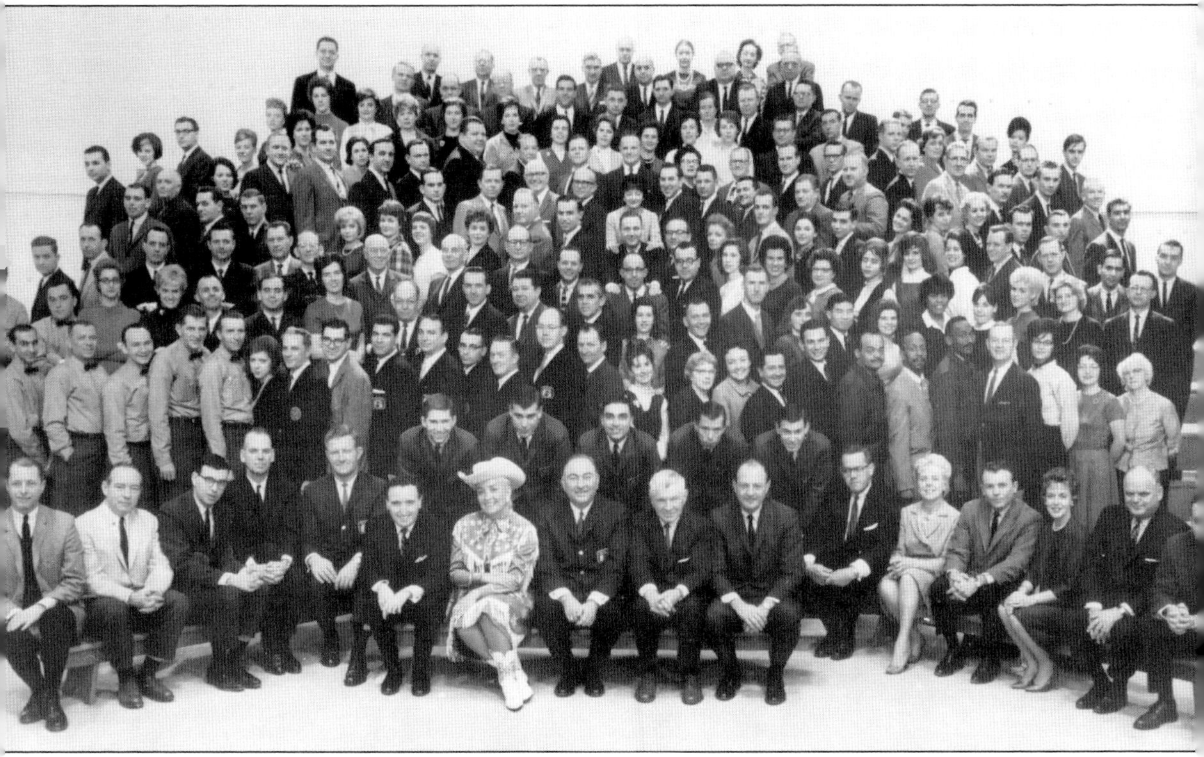

This team photograph shows everyone who worked at WFIL-TV in the mid-1960s. In the front row from left to right are organist and show host Larry Ferarri, children's show host Sally Starr, WFIL-TV owner Walter Annenberg, anchorman Gunnar Back, and meteorologist Francis Davis. (Special Collections Research Center, Temple University Libraries, Philadelphia, Pennsylvania.)

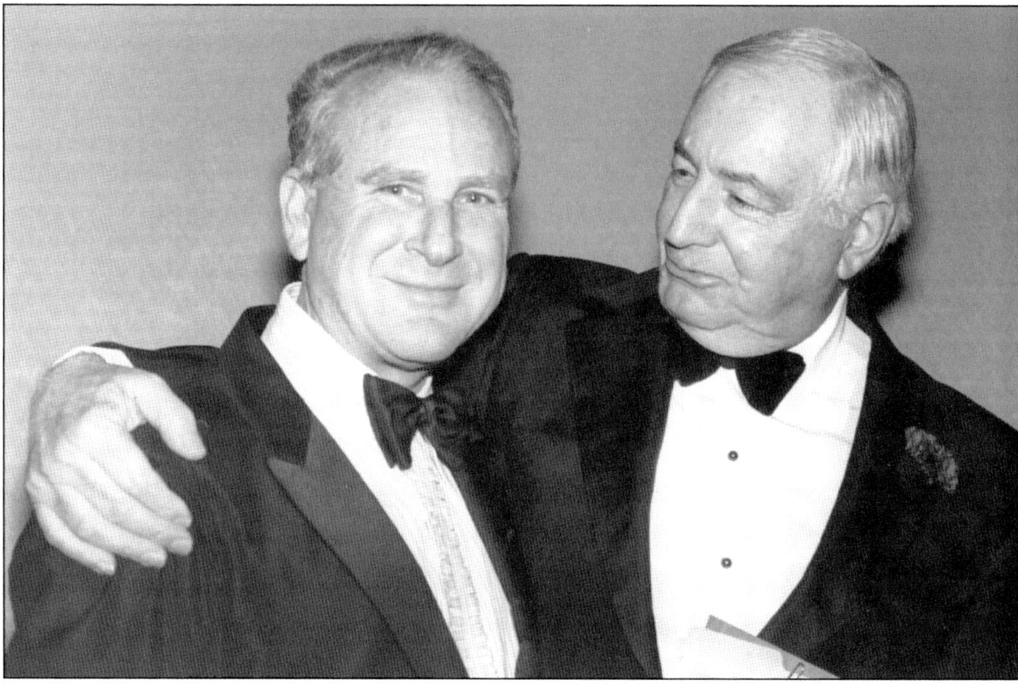

Lou Klein (left) and ambassador Walter Annenberg were keys to Channel 6's success from the 1940s to the 1970s. Annenberg owned the station until he sold it to Capital Cities Broadcasting in 1969.

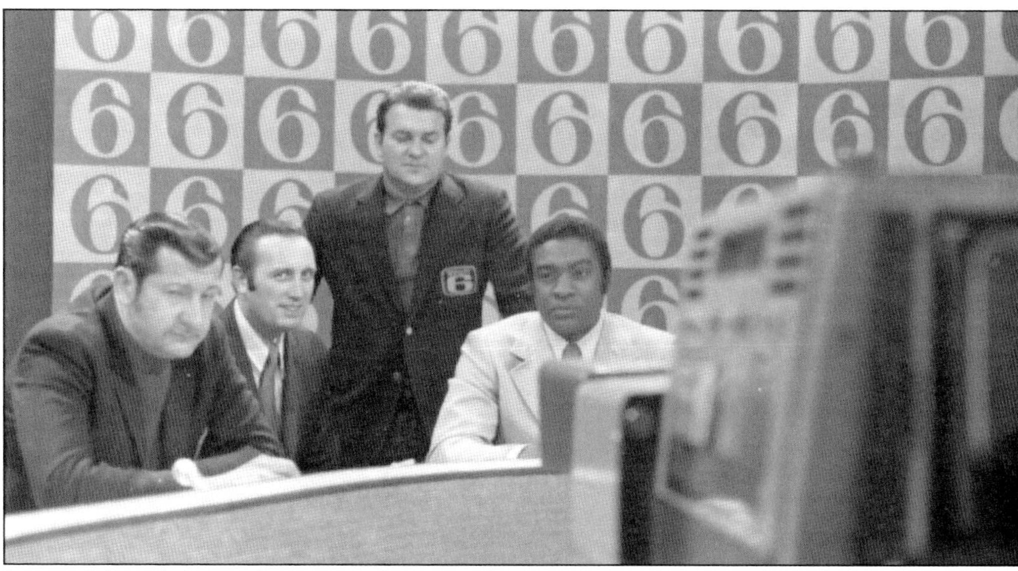

Pictured from left to right are Hank Latvin (studio camera operator), Bill Friel (studio camera operator and nationally known network football spotter), Tony Oscapinski (floor manager), and sportscaster Bill White on the *Action News* set in early 1970s. White was a former Phillies first baseman and longtime New York Yankees broadcaster. It is interesting that Oscapinski's blazer has the old WFIL logo, while the set has been changed to the new (and current) WPVI-TV Channel 6 logo. Former Channel 6 anchor Steve Levy says the studio technicians were major players in the early success of *Action News*. (WPVI-TV.)

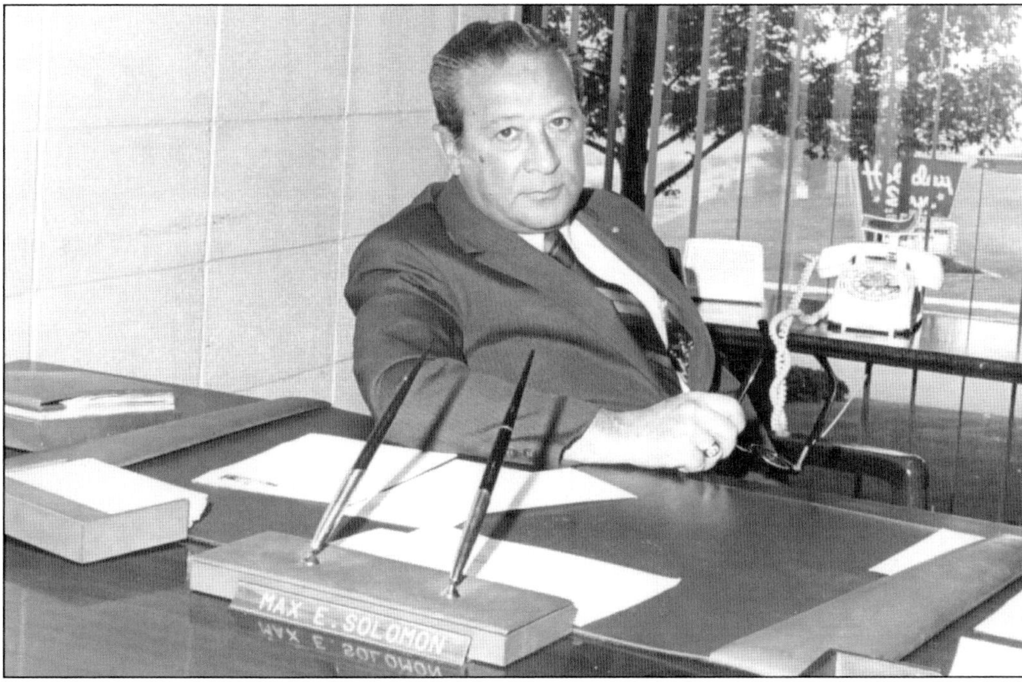

Salesman Max E. Solomon is pictured in his office at 4100 City Avenue. Notice the old Holiday Inn sign on the other side of Monument Road. (WPVI-TV)

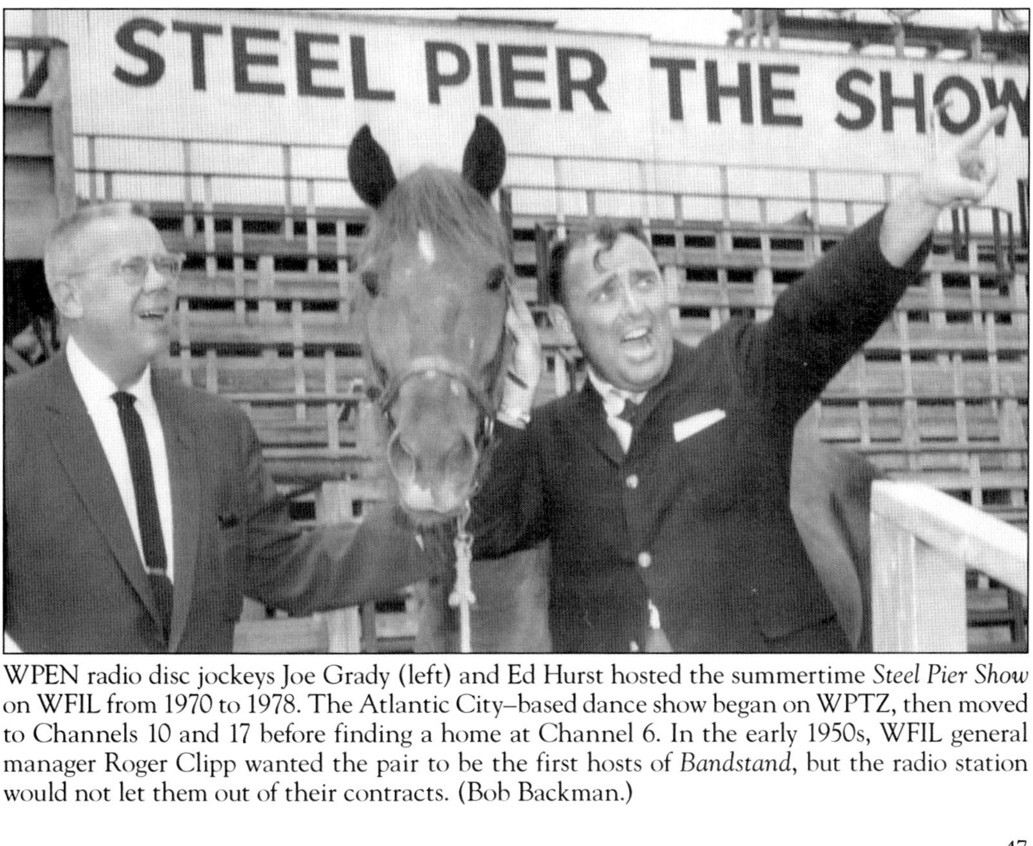

WPEN radio disc jockeys Joe Grady (left) and Ed Hurst hosted the summertime *Steel Pier Show* on WFIL from 1970 to 1978. The Atlantic City–based dance show began on WPTZ, then moved to Channels 10 and 17 before finding a home at Channel 6. In the early 1950s, WFIL general manager Roger Clipp wanted the pair to be the first hosts of *Bandstand*, but the radio station would not let them out of their contracts. (Bob Backman.)

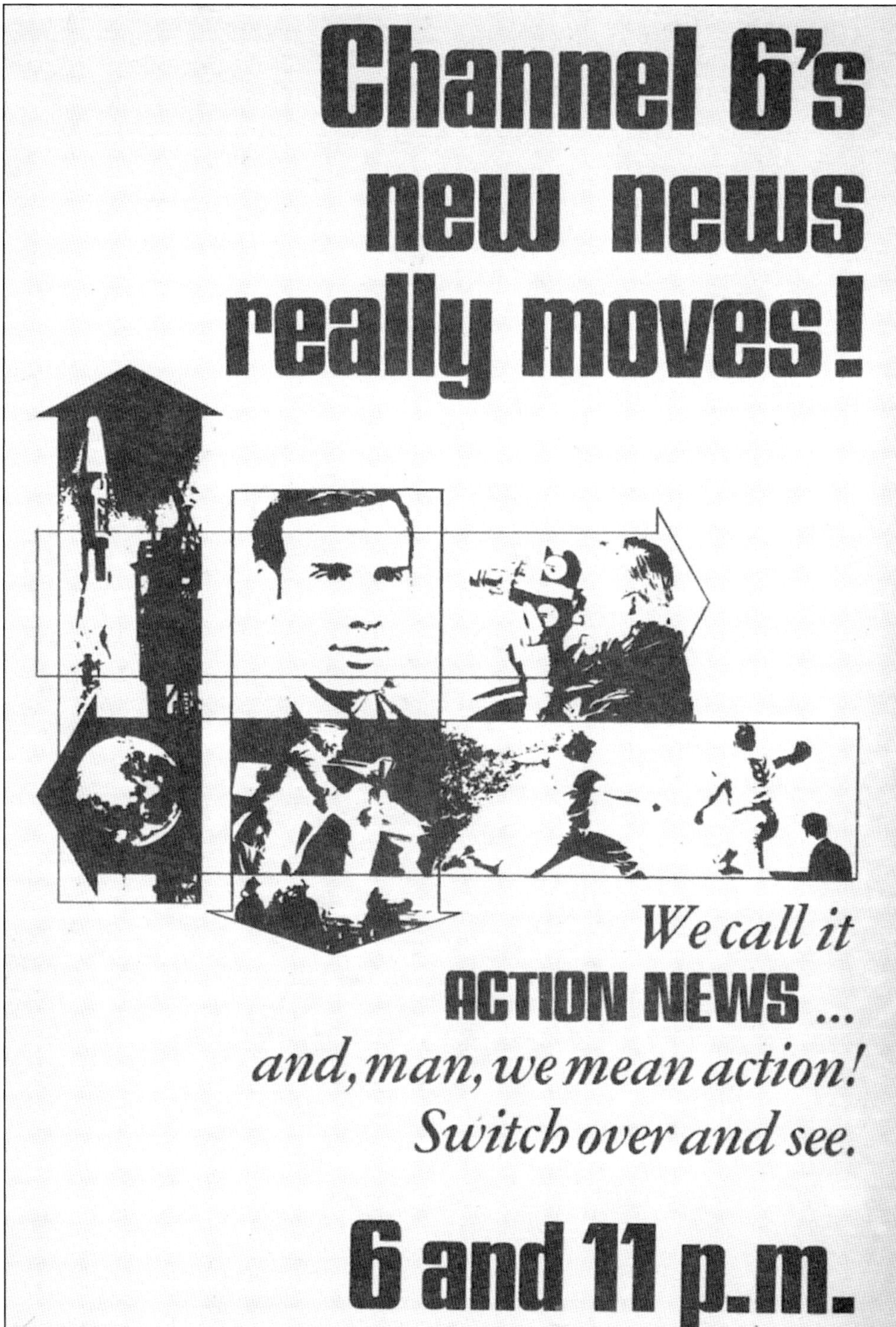

Channel 6 *Action News* debuted in April 1970. The fast-paced newscast was centered around anchorman Larry Kane and the idea that "Action News is Everywhere," shooting film for stories throughout the Delaware Valley. (WPVI-TV.)

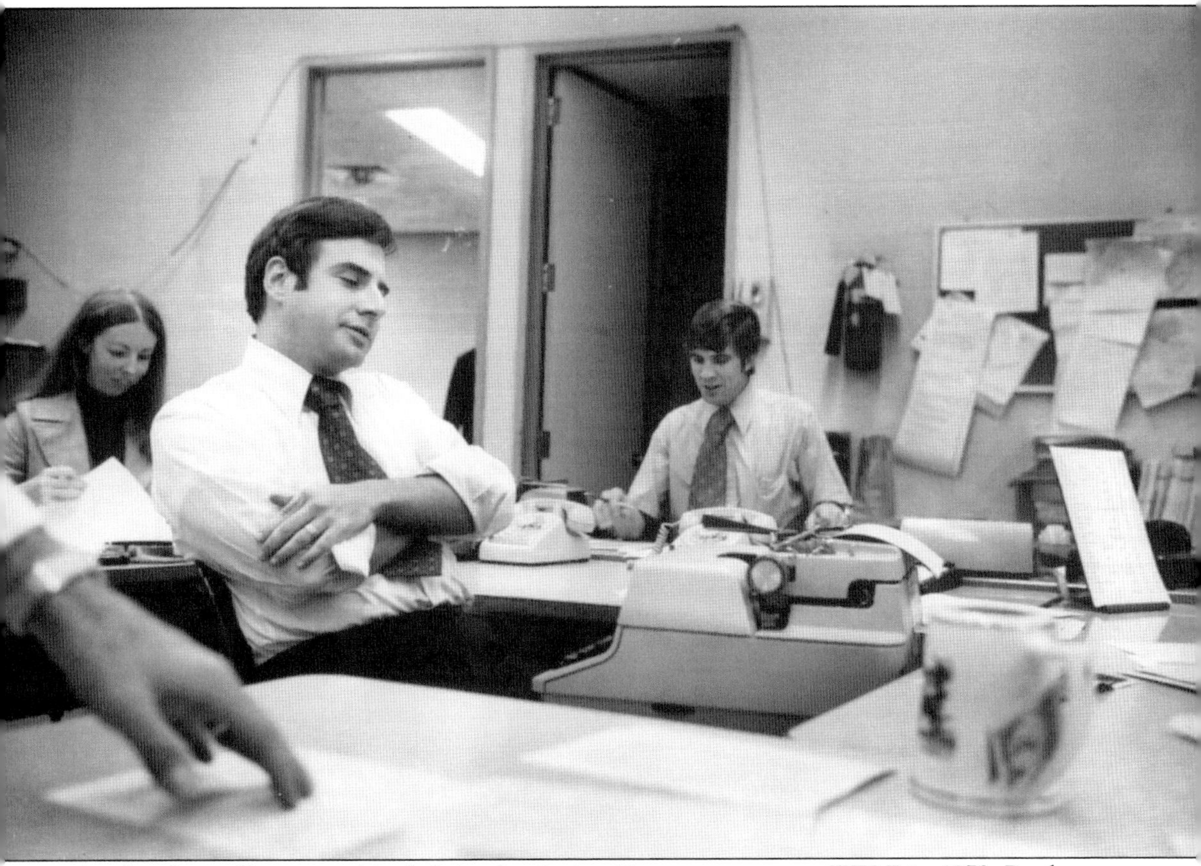

Anchor Larry Kane is at his desk in the *Action News* newsroom at WPVI in 1973. Producer Susan Durling is behind him, and an unidentified producer is seated to his left. Kane started as a news reporter at WFIL radio and then moved to TV in 1970. Kane had what was described as a fast-paced, "shotgun" style, in contrast to his more laid back competitors. (Special Collections Research Center, Temple University Libraries, Philadelphia, Pennsylvania.)

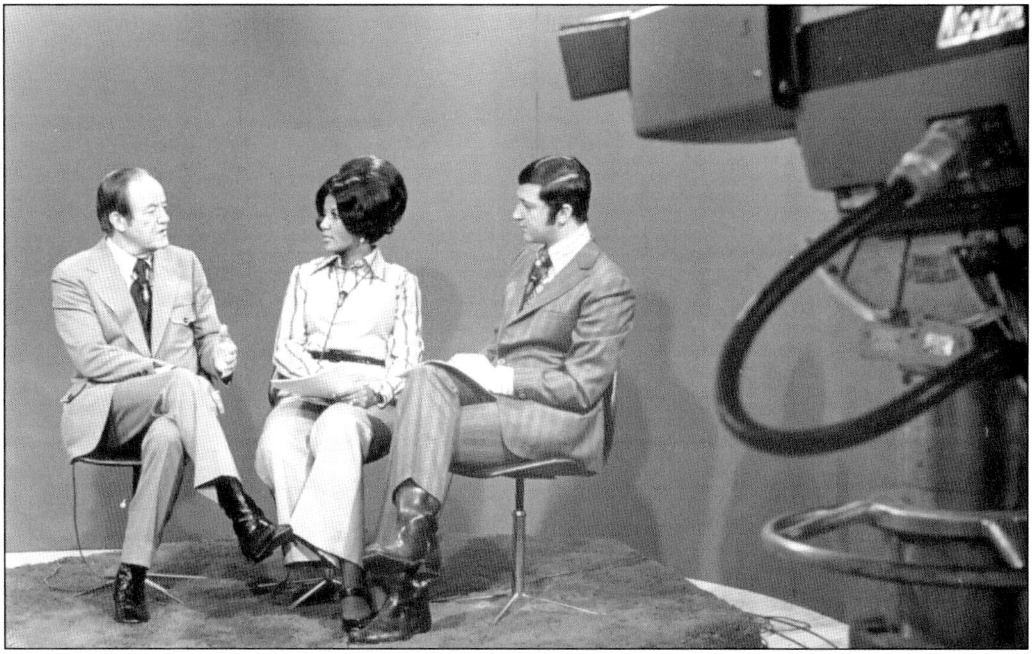

Mike Strug (right) and Diana Robinson (center) anchored *Action News at Noon* in the early 1970s. Robinson left for an anchor job in Detroit. Strug moved on to KYW and then WCAU. Here, they interview Sen. Hubert H. Humphrey. (WPVI-TV.)

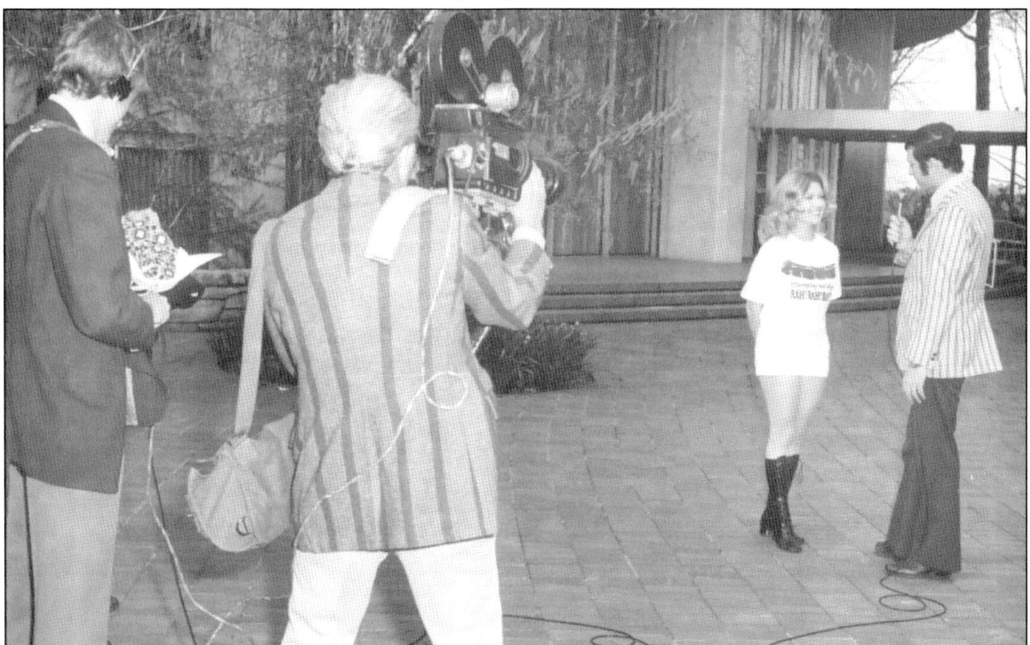

Even in the early 1970s, news was not all crashes and fires. Reporter Mike Strug interviews "Miss Plumping Hot Dog" outside the Channel 6 studios at 4100 City Avenue. The crew is cameraman Harold Hodgeman and soundman Bill Werndl. Werndl became a respected sports producer and talk show host. (WPVI-TV.)

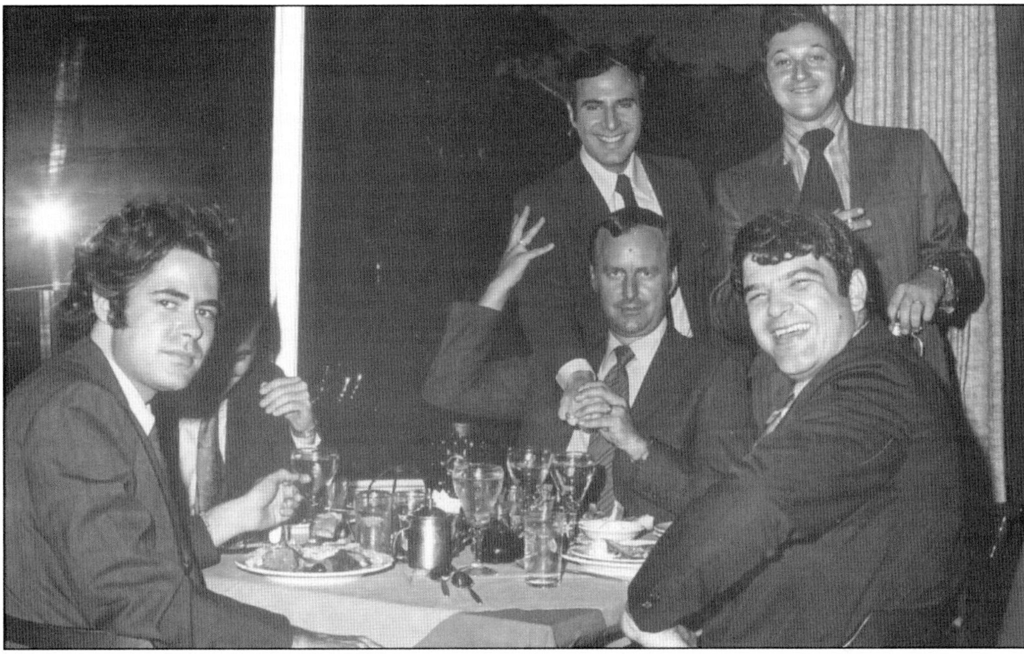

Channel 6 *Action News* became the top-rated newscast in the city in 1970. They celebrated at the nearby Bala Country Club. Standing from left to right are anchor Larry Kane and reporter Mike Strug. News director Mel Kampmann is seated with his hand on Kane. The two men in the foreground are said to be from the station's promotions department. (WPVI-TV.)

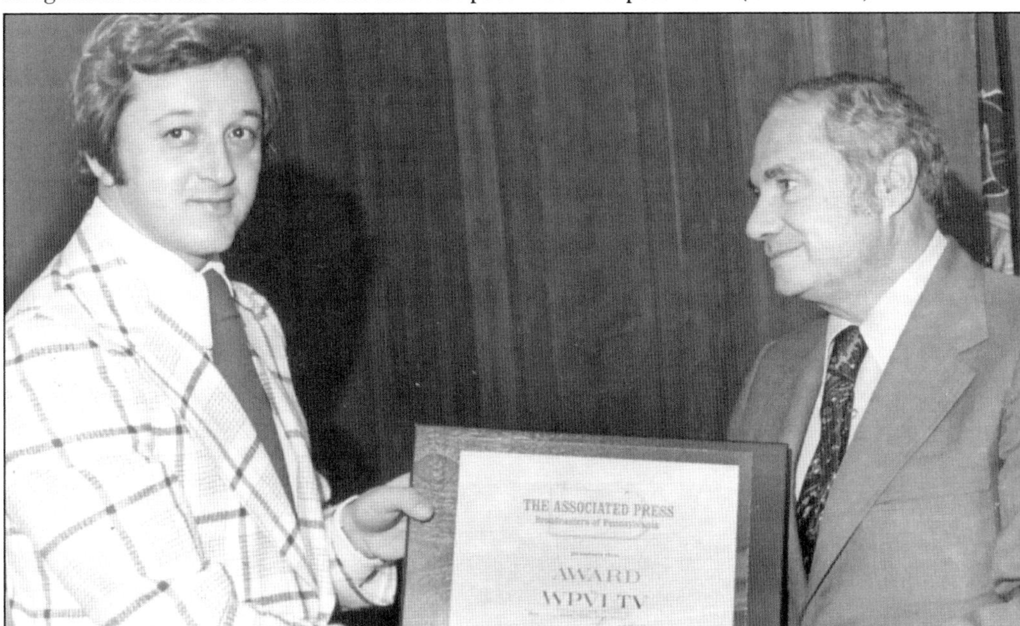

Pennsylvania governor Milton Shapp (right) presents Channel 6's Mike Strug with the Associated Press award for outstanding reporting. In 1970, Shapp accused WFIL and two other Pennsylvania TV stations owned by Triangle of a smear campaign against him. As a result, the FCC forced Triangle to sell its stations. (WPVI-TV.)

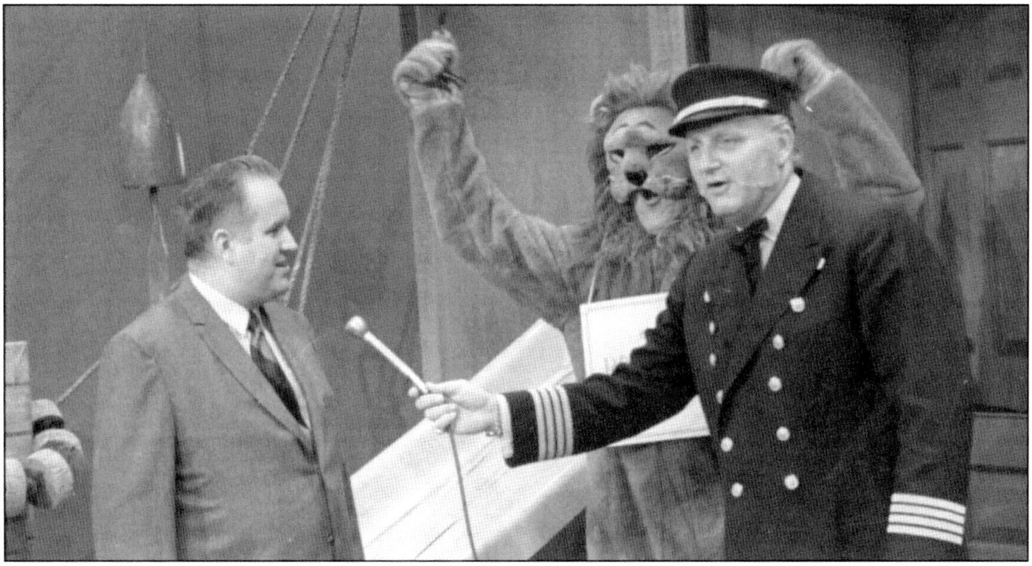

Carter Merbreir, known to all as Captain Noah, hosted his popular morning kids' show, *Captain Noah and his Magical Ark*, on WFIL/WPVI TV from 1967 to 1994. Noah and his real-life wife Patricia (also known as Mrs. Noah) welcomed guests aboard the "Magical Ark." Wayne Owens from General Electric is seen here announcing the winner of a contest to name the lion. The winning name was "Sir Roarsalot." The show received more than 20,000 entries from viewers. (Special Collections Research Center, Temple University Libraries, Philadelphia, Pennsylvania.)

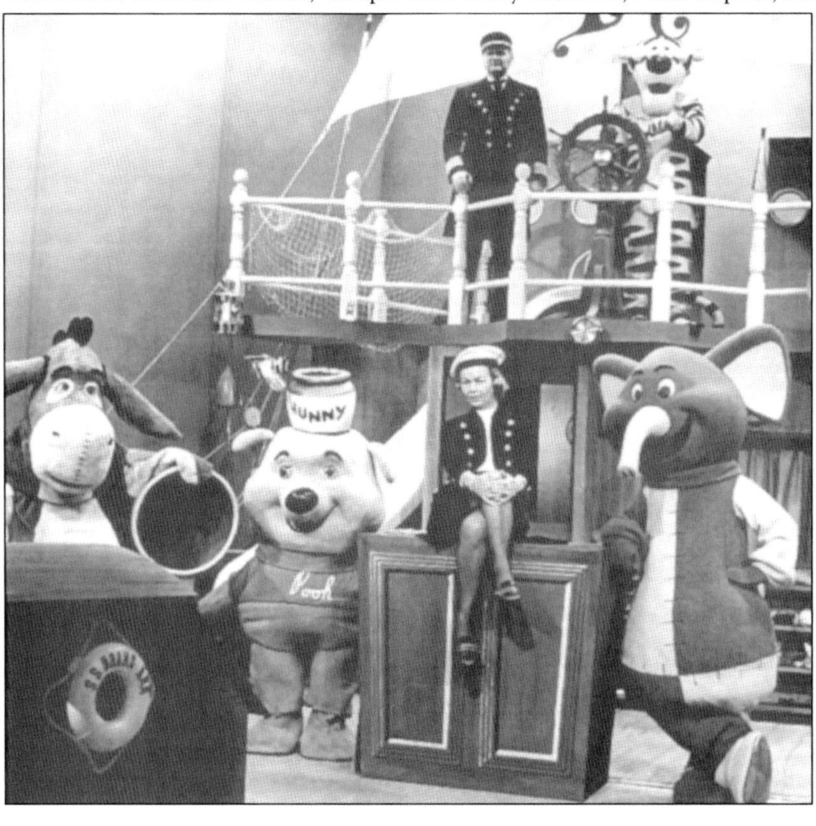

Mrs. Noah is pictured sitting amongst the characters of *Captain Noah and his Magical Ark*. (Special Collections Research Center, Temple University Libraries, Philadelphia, Pennsylvania.)

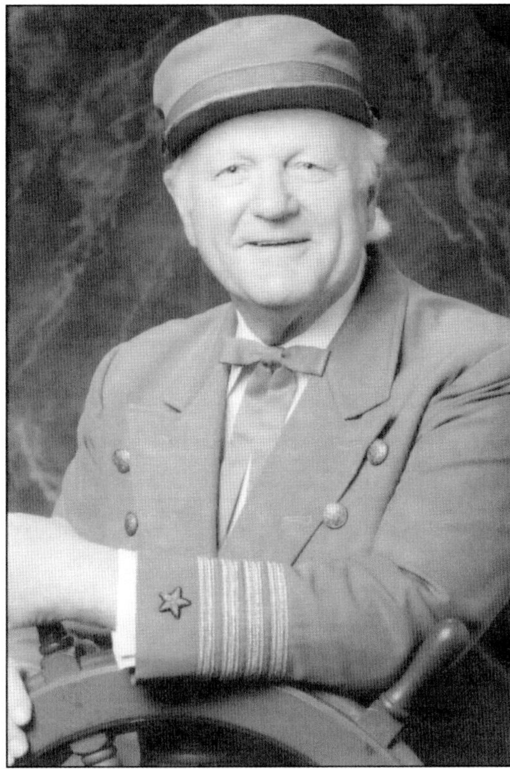

Seen here are promo cards for Captain and Mrs. Noah (Carter and Patricia Merbrier). They handed out the cards at their many public appearances around the Philadelphia area. Captain Noah was an ordained Lutheran minister and former Philadelphia police chaplain. (Both, WPVI-TV.)

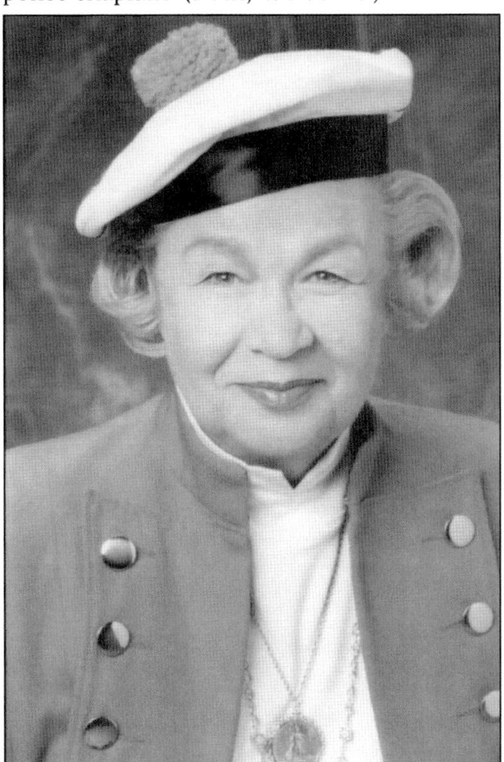

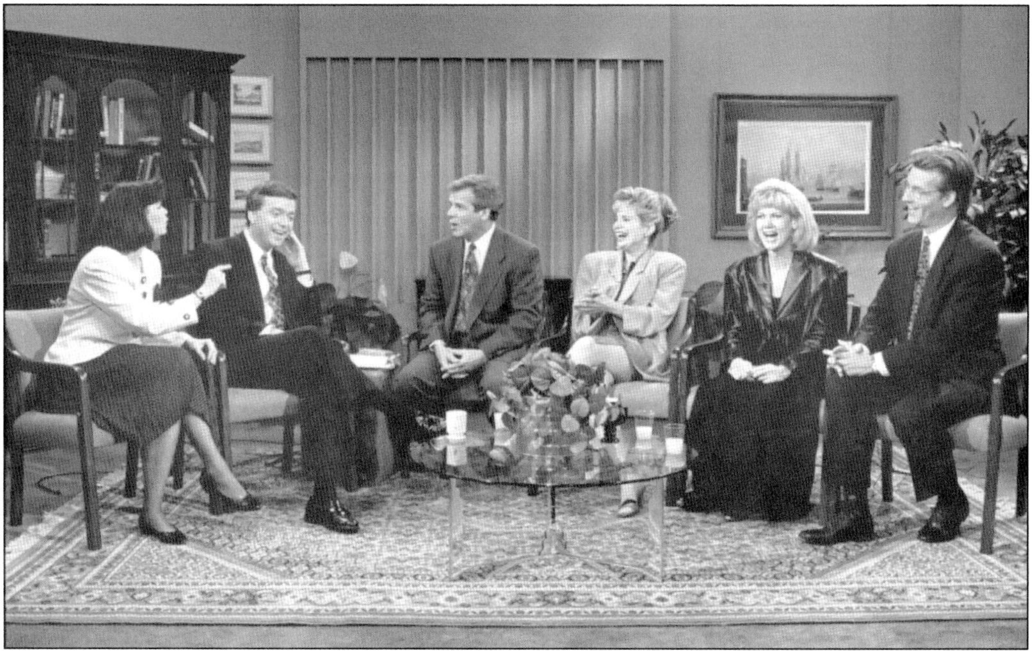

WPVI's Morning talk show AM *Philadelphia* celebrated its 15th anniversary in 1993 with a reunion of all of its hosts. From left to right are hosts Liz Starr and Wally Kennedy, and previous hosts Dave Roberts, Claire Carter, Janet Davies, and Harry Martin. (WPVI-TV.)

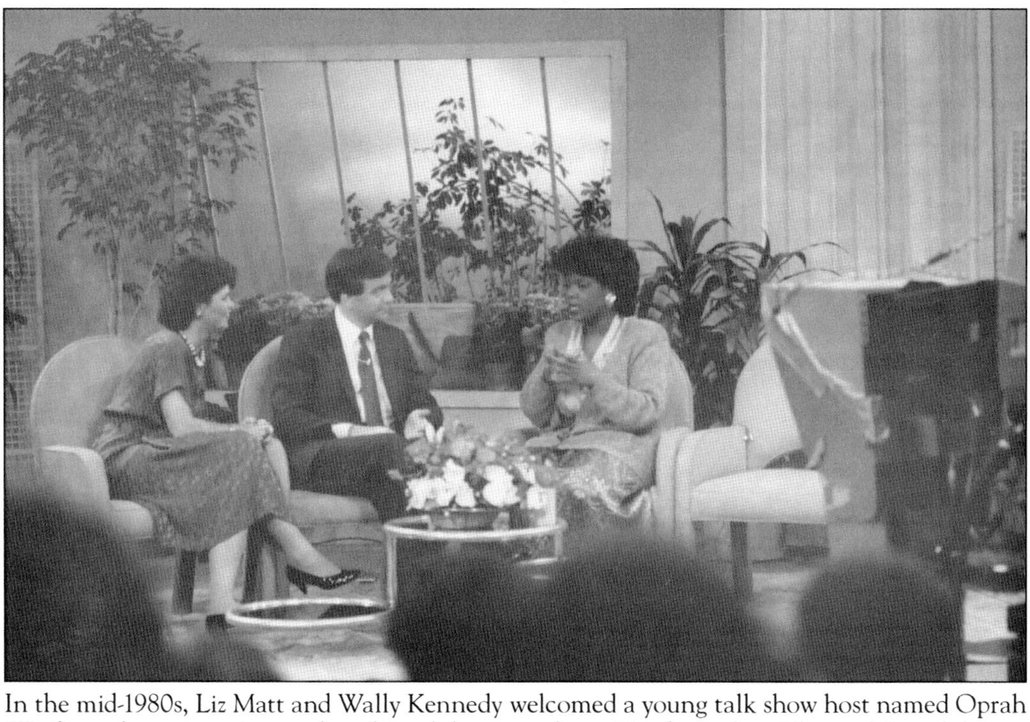

In the mid-1980s, Liz Matt and Wally Kennedy welcomed a young talk show host named Oprah Winfrey, who was getting ready to launch her own show. *Oprah* aired on Channel 6 weekdays at 4:00 p.m. until 2011. With Oprah as a lead in, *Action News at 5* was the top local news program in its time slot for decades. (WPVI-TV.)

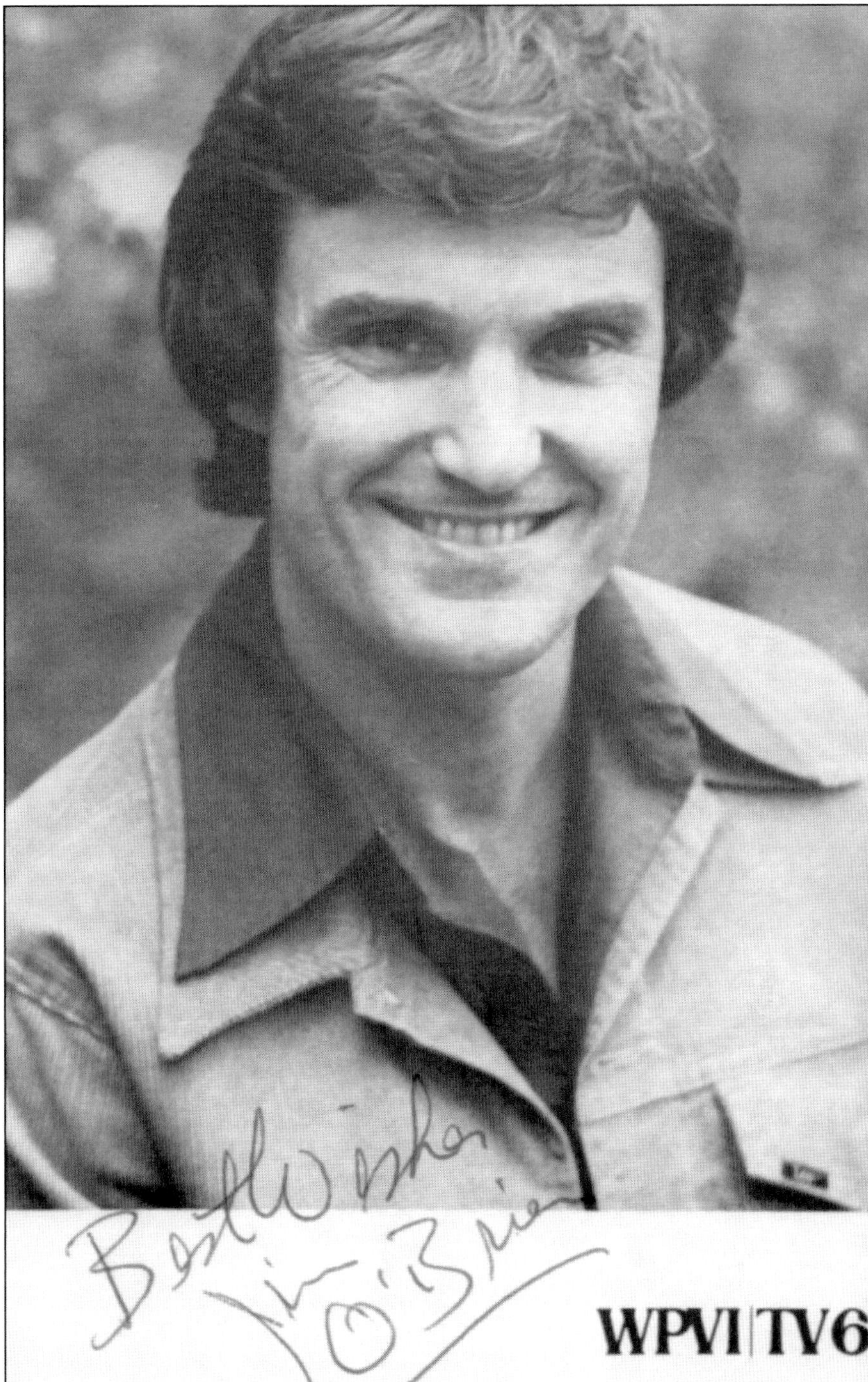

Jim O'Brien was arguably the most popular personality in Philadelphia TV history. O'Brien started as a disc jockey on WFIL radio before becoming the weatherman on *Action News*. He later anchored *Action News at 5* and hosted the popular *Dialing For Dollars* in the morning. The Texas native was known for his down-home delivery and weather map with happy- and sad-face clouds. O'Brien was killed in a skydiving accident in Montgomery County in September 1983. (WPVI-TV.)

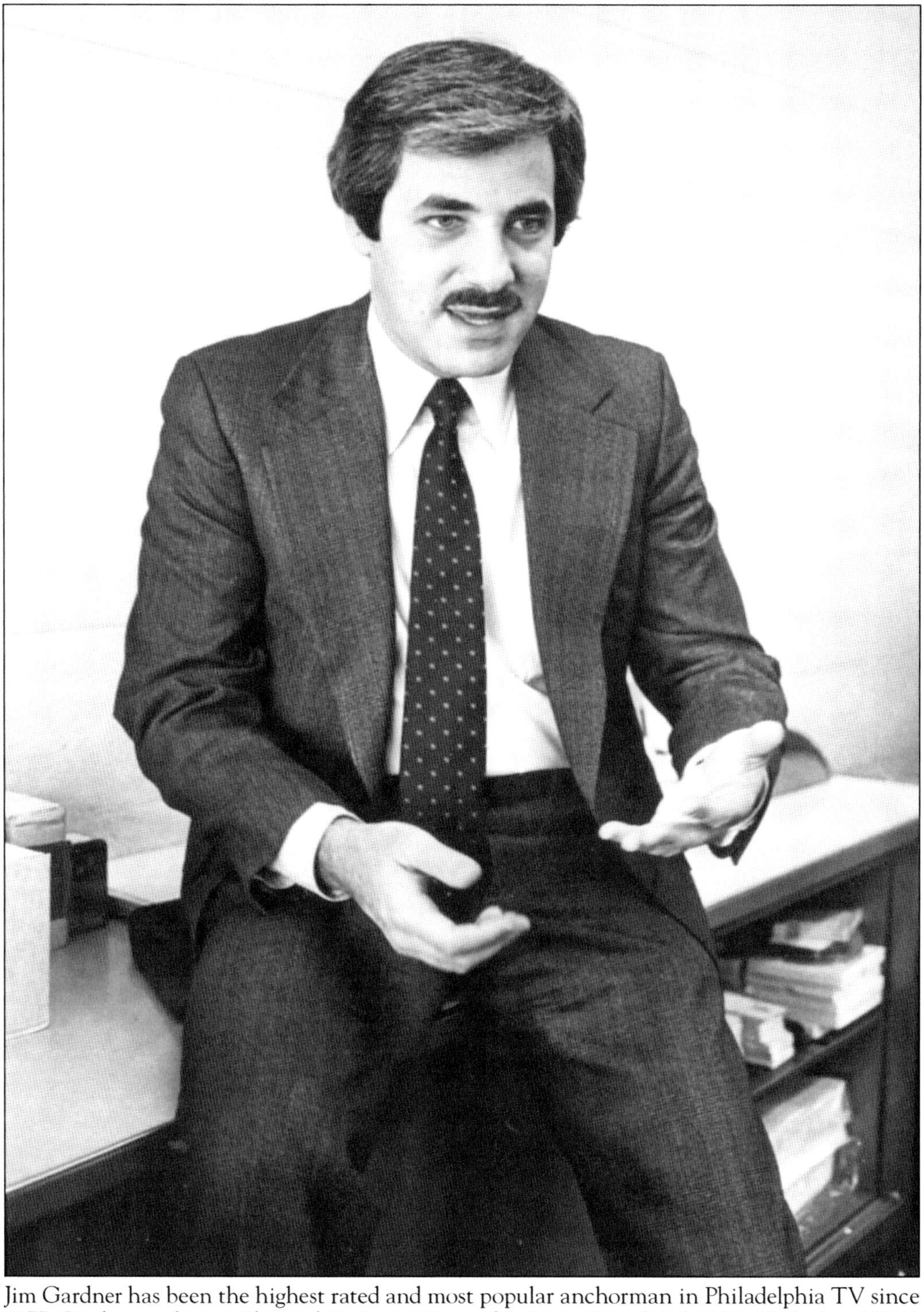

Jim Gardner has been the highest rated and most popular anchorman in Philadelphia TV since 1977. Gardner took over Channel 6's *Action News* after Larry Kane left for New York. As of 2015, Gardner was the dean of local TV anchormen, still on top at 6:00 p.m. and 11:00 p.m. (Special Collections Research Center, Temple University Libraries, Philadelphia, Pennsylvania.)

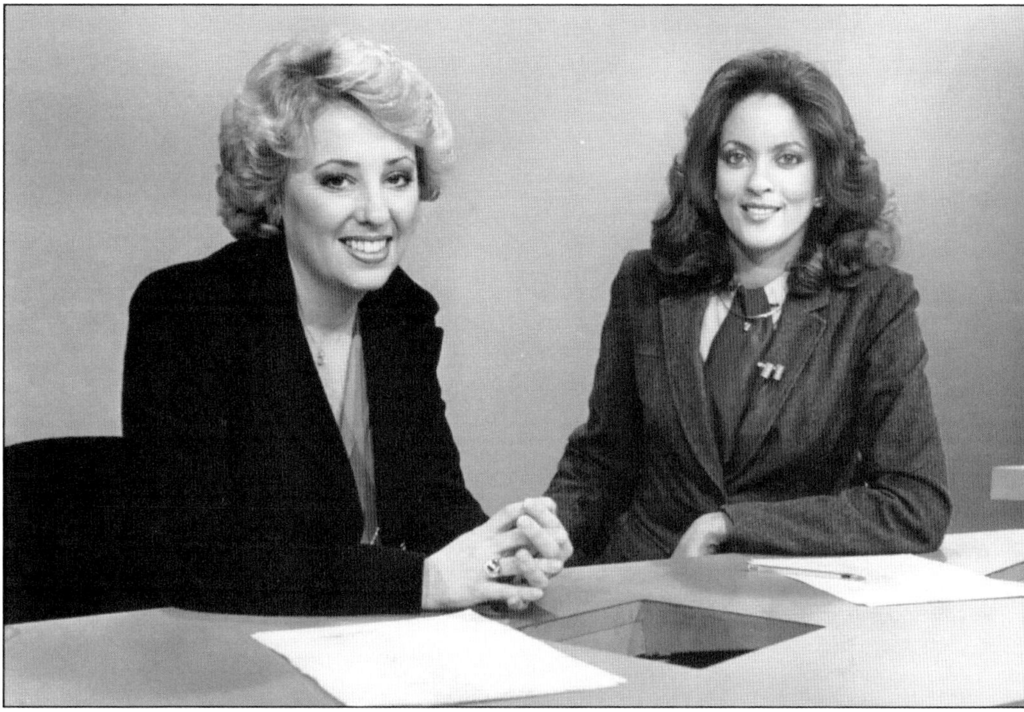

Chris Wagner and Lisa Thomas-Laury anchored *Action News at Noon* in the 1980s and 1990s. (Special Collections Research Center, Temple University Libraries, Philadelphia, Pennsylvania.)

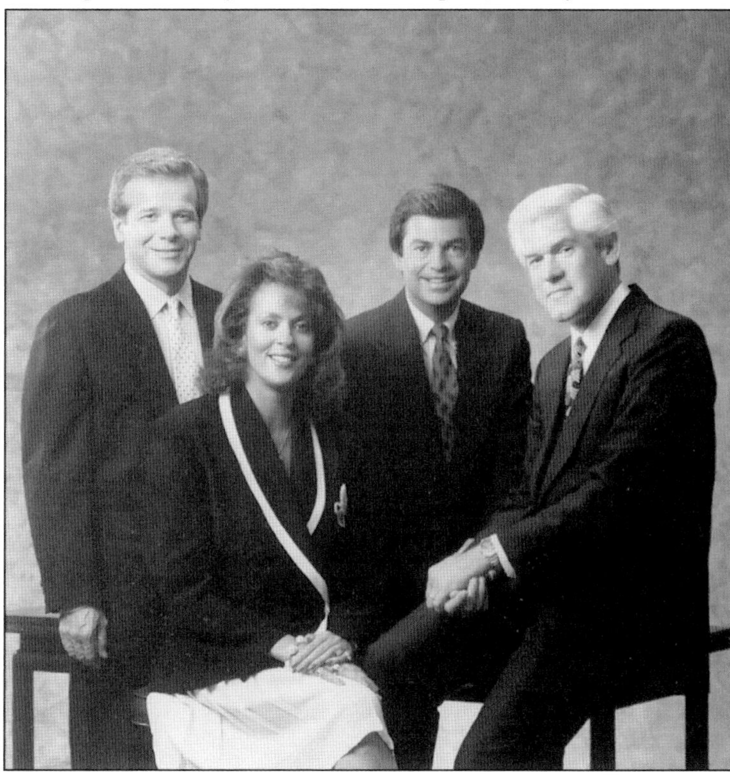

The team from *Action News at 5* in the 1990s included, from left to right, weatherman Dave Roberts (father of actor David Boreanaz), anchor Lisa Thomas-Laury (an anchor at Channel 6 since 1978), sportscaster Scott Palmer (now vice president of communications for the Phillies), and anchor Marc Howard, who left for KYW in 2003. (WPVI-TV.)

ACTION NEWS FOR 6PM XXXXXX TUESDAY OCT. 19,1976

SD	34	BREAK #4 TITLES		VTR	1:33	6:00:00
LK	34	TEASE		LV/KEY-LK	:10	6:00:10
DT	35	MURDER ARRAIGN	(MUR. INVEST)	CHR/SL/FM/CT/KEY-DT#	1:40	6:01:50
LK	36	NARBERTH MURDERER		LV	:15	6:02:05
CV	37	BUS AX	(VC)	CHR/VC	:30	6:03:35
CG	38	PENNSBURY	(T.(STRIKE)	CHR/SL	:15	6:03:50
CG	39	CHERRY HILL		FM	:25	6:04:15
CG	40	OTHERS/PROTEST	(T.STRIKE)	CHR/SL/FM	:25	6:04:40
JG	41	SUB TEACHERS		LV/FM/CT/KEY-JG	1:30	6:06:10
LK	42	FELDMAN		LV	:15	6:06:25
SD	43	COML #5	(BUMPER)	SL/VTR	1:40	6:08:05
CG	44	CHARTER HASSLE	(NJ POLITIX)	CHR/SL/FM/KEY-LK"	:25	6:08:30
VN	45	PROFILE	(4TH CONG)	CHR/SL/FM/CT/KEY-VN	2:25	6:10:55
AVP	46	PROMO	(PROFILE)	CHR/SL	:10	6:11:05
CV	47	TWO GUYS		LV/FM	:25	6:11:30
LK	48	NJ OFFICE		LV	:15	6:11:45
SD	49	COML #6	(BUMPER)	SL/VTR	2:10	6:13:55
LK	50	REGIONAL ENERGY		LV/KEY-LK	:15	6:14:05
HS	51	BANKS	(BANKS)	CHR/SL/VC/CT/KEY-HS*	1:40	6:15:50
SD	52	SPORTS	(SPORTS)	CHR/SL/FM/VC/VTR/KEY-JP#	3:30	6:19:35
CG	53	X-MAS PKGS		LV/VC	:25	6:20:00
SD	54	COML #7	(WIDESHOT)	VTR/KEY-ACCU	1:40	6:21:40
LK	55	LOCAL WRAP	(WRAP)	CHR/SL/FM/SL(Amb Ax)	:50	6:22:30
DT	56	PUMPKINS		FM	:25	6:22:45
SD	57	WEATHER		LV/JIM/KEY-JO	2:30	6:25:15
JB	58	ROLLS ROYCE		LV/VC/CT/KEY-JB*	1:30	6:26:45
SD	59	BYE		3-SHOT/FM/CT/KEYS	2:10	6:28:55

This is a typical show rundown from Channel 6 *Action News at 6* for October 19, 1976. Each column means something for the producer, director, studio crew, and anchor Larry Kane. (Author's collection.)

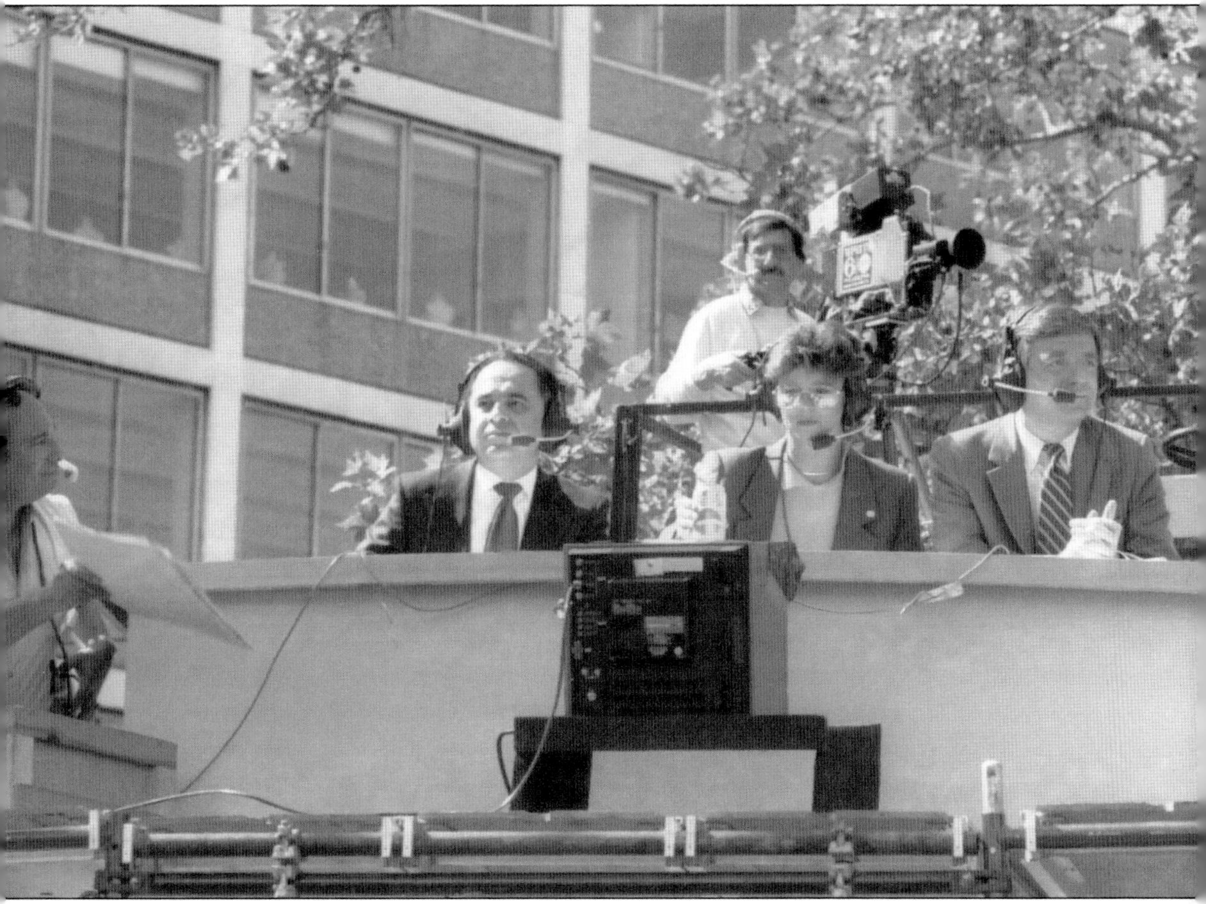

WFIL/WPVI has always been the leader in local parade coverage. This image shows (from left to right) cohosts Diego Castiano, Liz Matt, and Wally Kennedy at Philadelphia's Puerto Rican Day parade in the 1980s. Channel 6 has also broadcast the local Thanksgiving Day parade for decades. (WPVI-TV.)

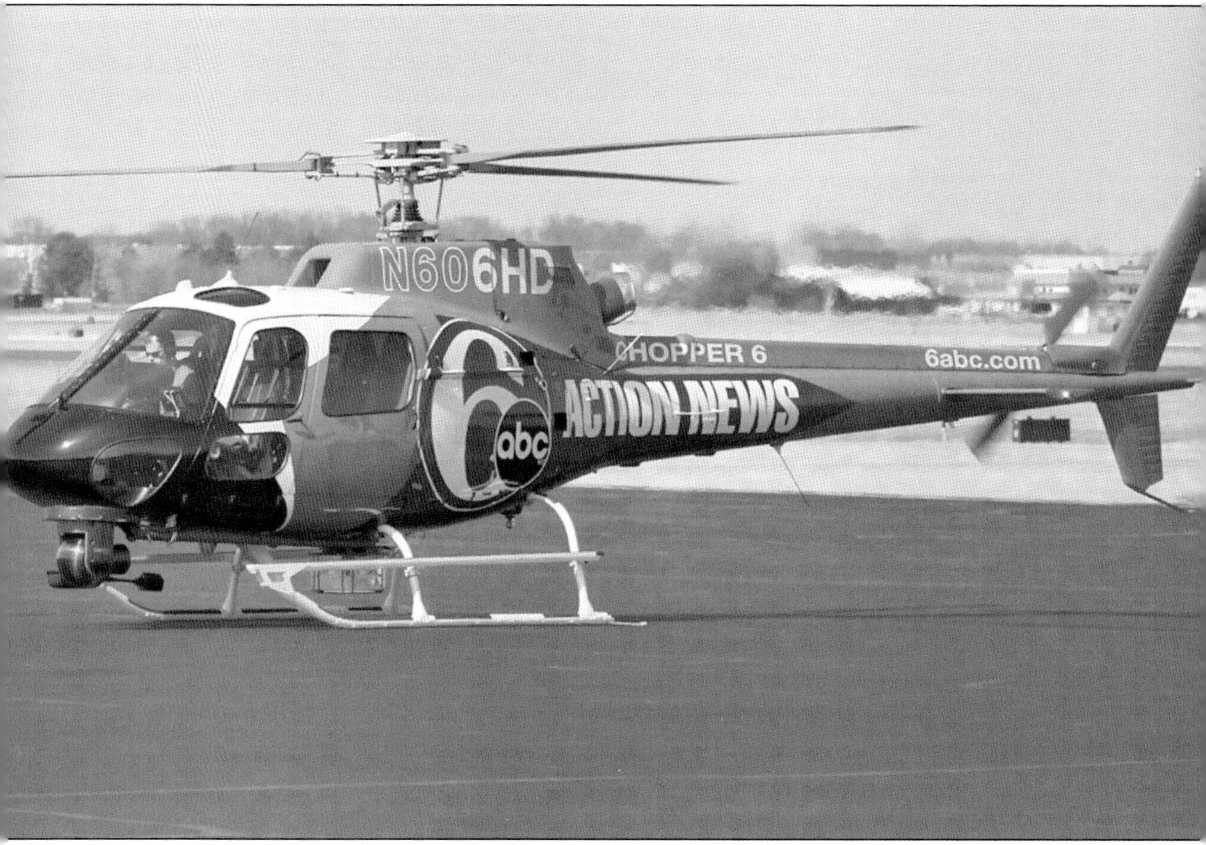

Chopper 6 has been a fixture in the sky over the Delaware Valley for more than 30 years. The current aircraft captures video in high definition. It is seen here at its home base, Northeast Philadelphia Airport. (Author's collection.)

Three

WCAU Channel 10

WCAU–TV started broadcasting from the WCAU radio building at 1633 Chestnut Street in Center City in 1948. The station was owned by the *Philadelphia Bulletin* newspaper. In 1952, The *Bulletin* moved the station to a brand new building at City Avenue and Monument Road in Bala Cynwyd. WCAU has always claimed the building was the first ever designed as a broadcast center for TV and radio. (WFIL also makes the claim for its Forty-sixth and Market Streets building). The building's multiple studios and large outdoor space allowed WCAU to produce groundbreaking programs. One of those shows was *Action in the Afternoon*, a daily, live Western drama produced in the station's back lot. Jane Norman's *Pixanne* and Gene London's *Cartoon Corners* were two of the iconic WCAU children's programs. Gene Crane hosted *Candy Carnival*, *Grand Chance Roundup*, and *Mr. & Mrs.* (with his real wife). Crane worked at WCAU-TV for almost 60 years. He also did the weather on Channel 10 News. Edie Huggins was a reporter and talk show host at the station for 40 years. John Facenda was the face of WCAU news from 1948 to 1973. His style and voice were legendary. He worked alongside Edie Huggins, who was a reporter and talk show host at the station for 40 years. Many other famous faces worked at WCAU in the course of their careers: Ed McMahon, Tom Brookshier, Jack Whitaker, Brian Williams, Matt Lauer (cohost of the short-lived talk show *Live on City Line* with Sheela Allen-Stephens), NBC News White House correspondent Kristen Welker, and former *CBS Morning Show* host Jane Robelot. CBS owned WCAU from 1958 to 1995, when the station was sold to NBC as part of a larger affiliate switch.

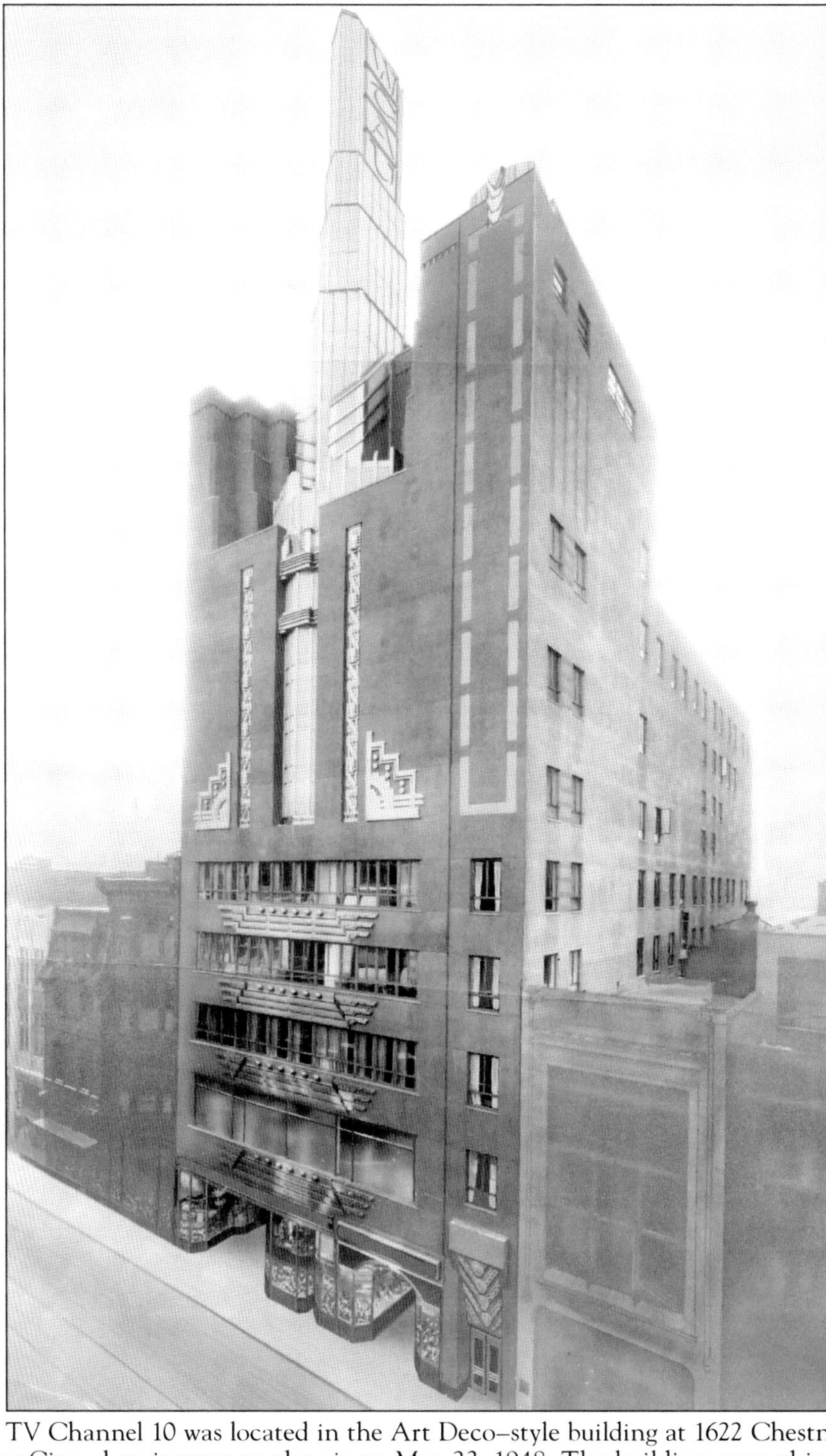

WCAU TV Channel 10 was located in the Art Deco–style building at 1622 Chestnut Street in Center City when it went on the air on May 23, 1948. The building opened in 1933 for WCAU radio. (WCAU-TV/Special Collections Research Center, Temple University Libraries, Philadelphia, Pennsylvania.)

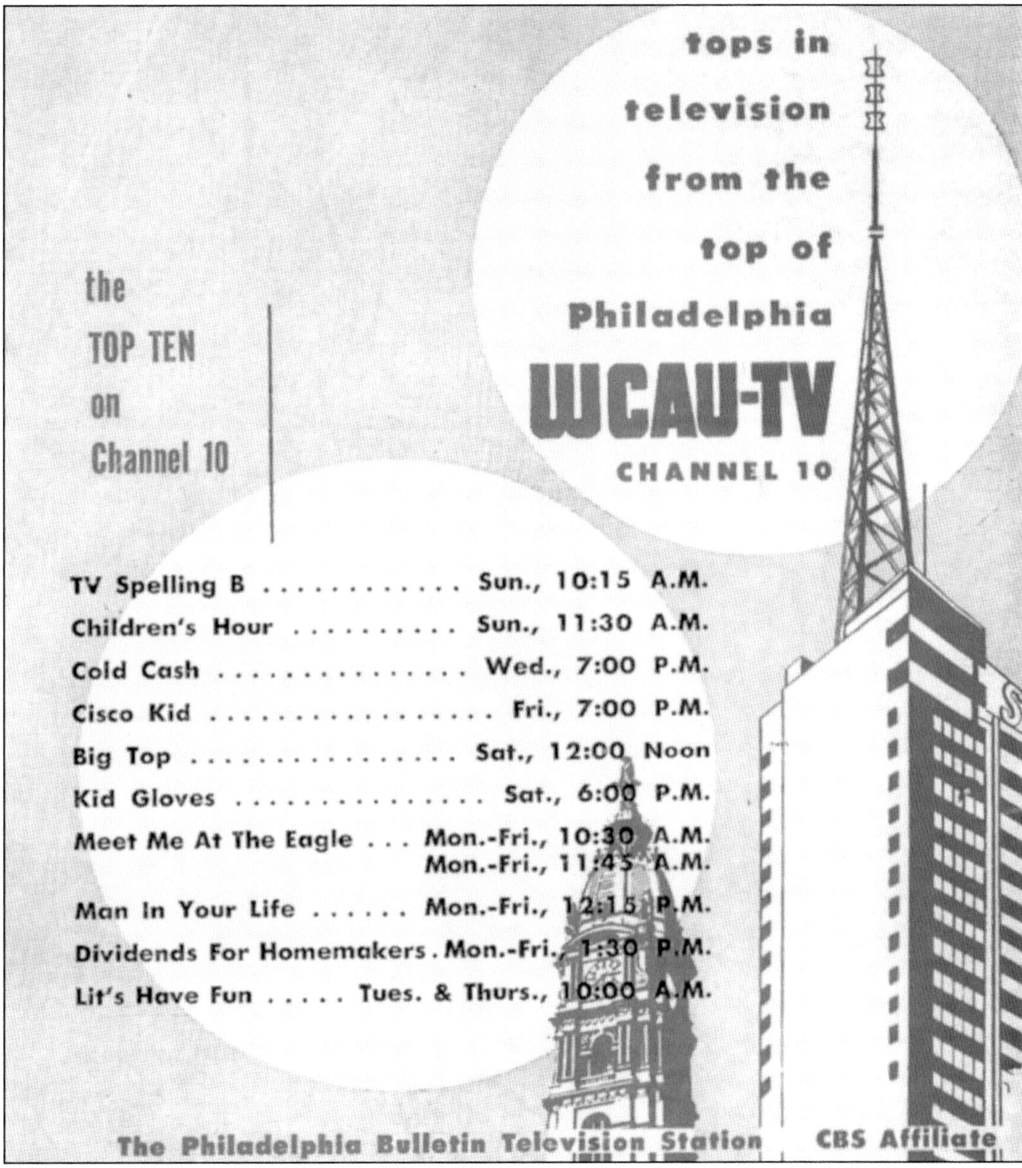

WCAU-TV's tower was first located on top of the PSFS Building at Twelfth and Market Streets. The building was the tallest skyscraper in the city when the station went on the air in 1948. A plan to move the station there never materialized. This promotional ad shows the station's daily lineup in June 1950. (WCAU-TV.)

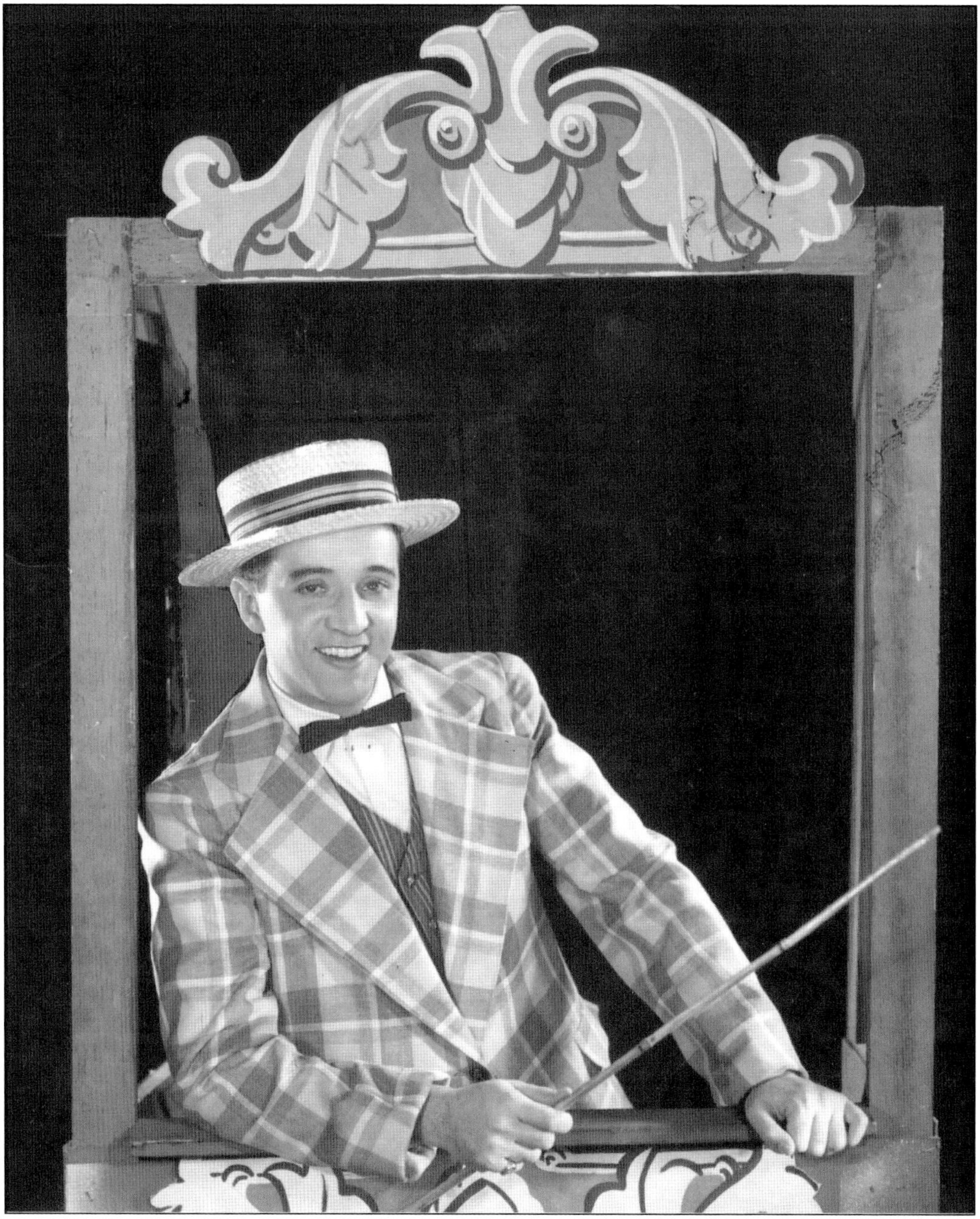

Gene Crane dressed as a carnival barker for one of the many programs he hosted at WCAU. His shows include *Candy Carnival*, *Grand Chance Roundup*, and *Mr. & Mrs*, which he cohosted with his real-life wife, Joan. Crane's son David was one of the producers of the sitcom *Friends*. (WCAU-TV/ Special Collections Research Center, Temple University Libraries, Philadelphia, Pennsylvania.)

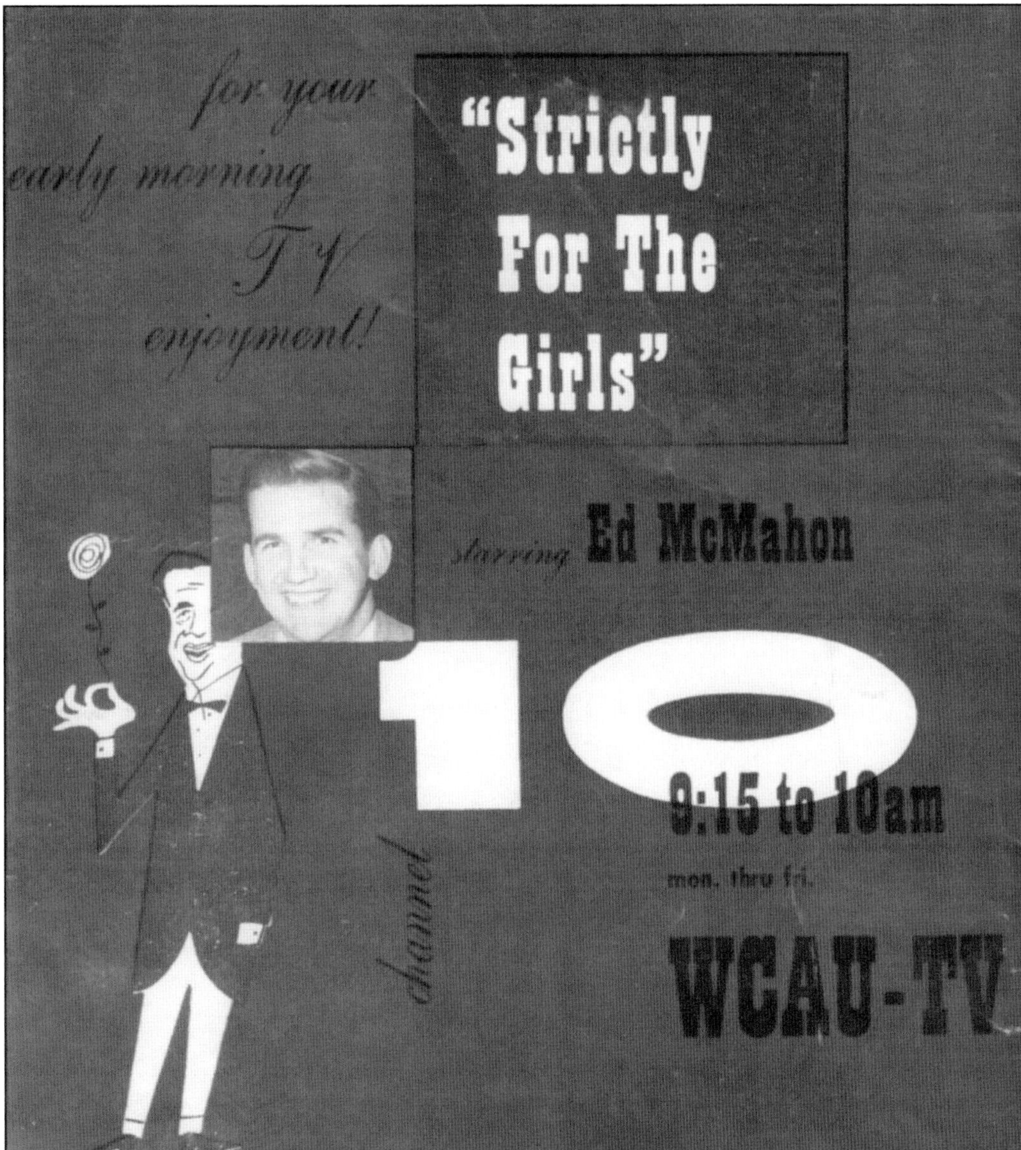

Ed McMahon got his start in television at WCAU. In 1950, he hosted a morning program called *Strictly For the Girls*. McMahon also played a clown on the talent show *Big Top*. He later became a household name as Johnny Carson's sidekick on NBC's *Tonight Show* and as host of *Star Search*. During his time at Channel 10, McMahon lived next door to *Bandstand* host Dick Clark. (WCAU-TV.)

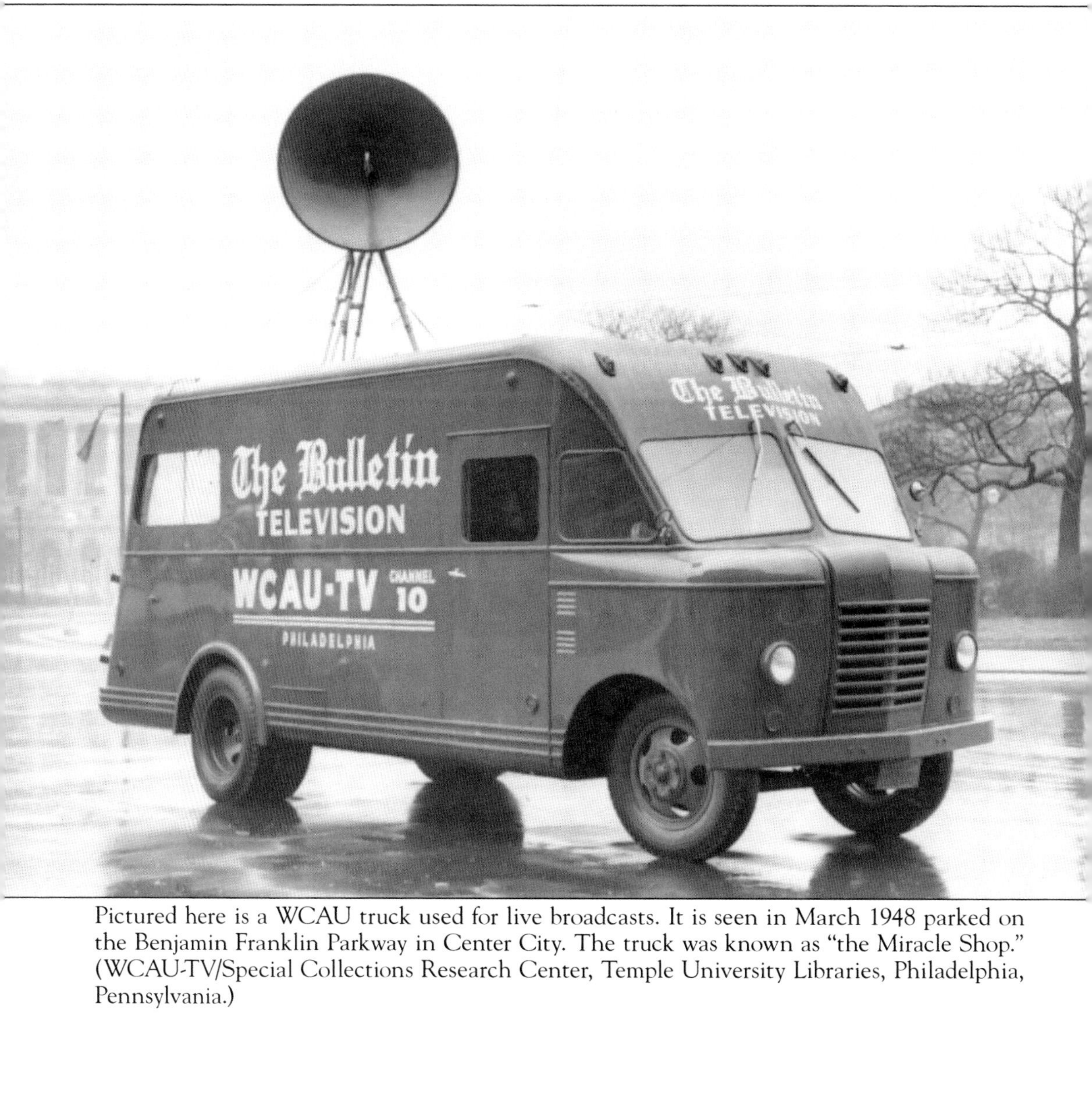

Pictured here is a WCAU truck used for live broadcasts. It is seen in March 1948 parked on the Benjamin Franklin Parkway in Center City. The truck was known as "the Miracle Shop." (WCAU-TV/Special Collections Research Center, Temple University Libraries, Philadelphia, Pennsylvania.)

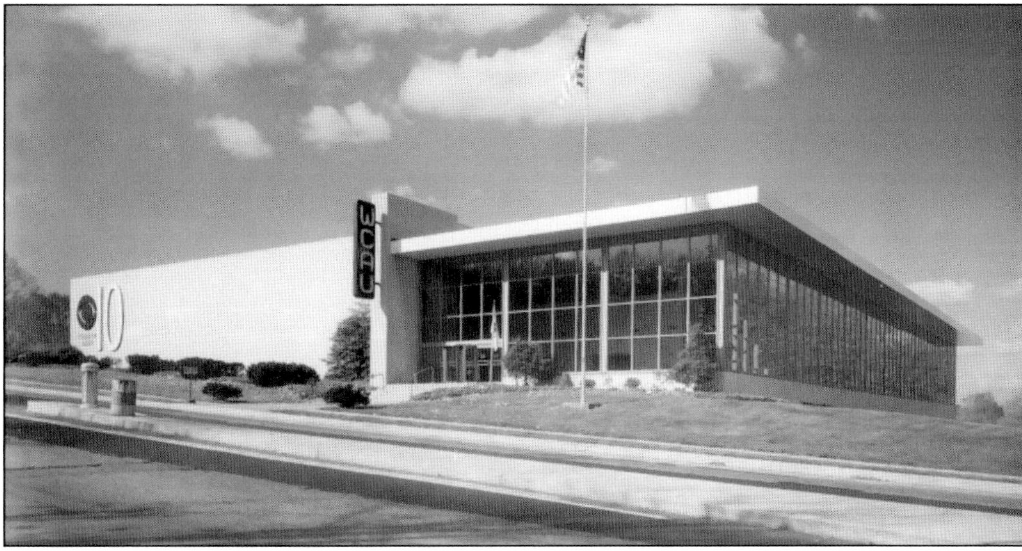

In 1952, The *Philadelphia Bulletin* moved WCAU-TV and Radio to a state-of-the-art building designed by architects George Howe and Robert Montgomery Brown. The building is at the corner of City Avenue and Monument Road in Bala Cynwyd, Montgomery County. WCAU claimed it was the first-ever structure designed to be a broadcasting facility (TV and radio) from the ground up. (Lower Merion Historical Society.)

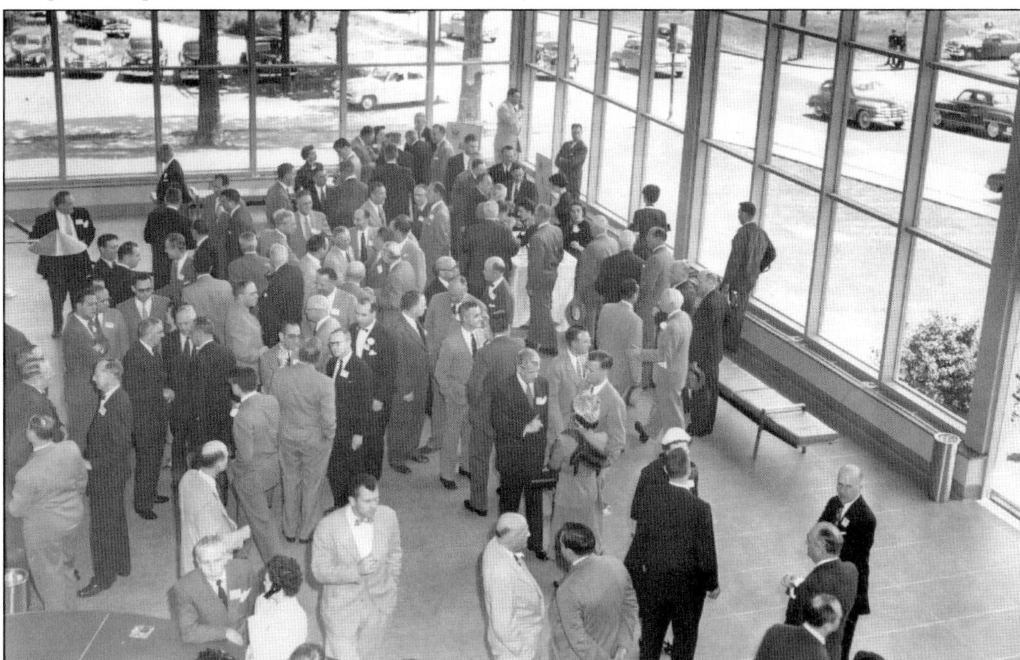

Dignitaries and guests filled the lobby on Dedication Day for WCAU-TV's new building, May 27, 1952. Over the years, this area served as the newsroom, sales offices, and a studio for the NBC 10 *10! Show.* The empty ground on the opposite corner was a golf course; it is now a shopping center and hotel. (WCAU-TV/Special Collections Research Center, Temple University Libraries, Philadelphia, Pennsylvania.)

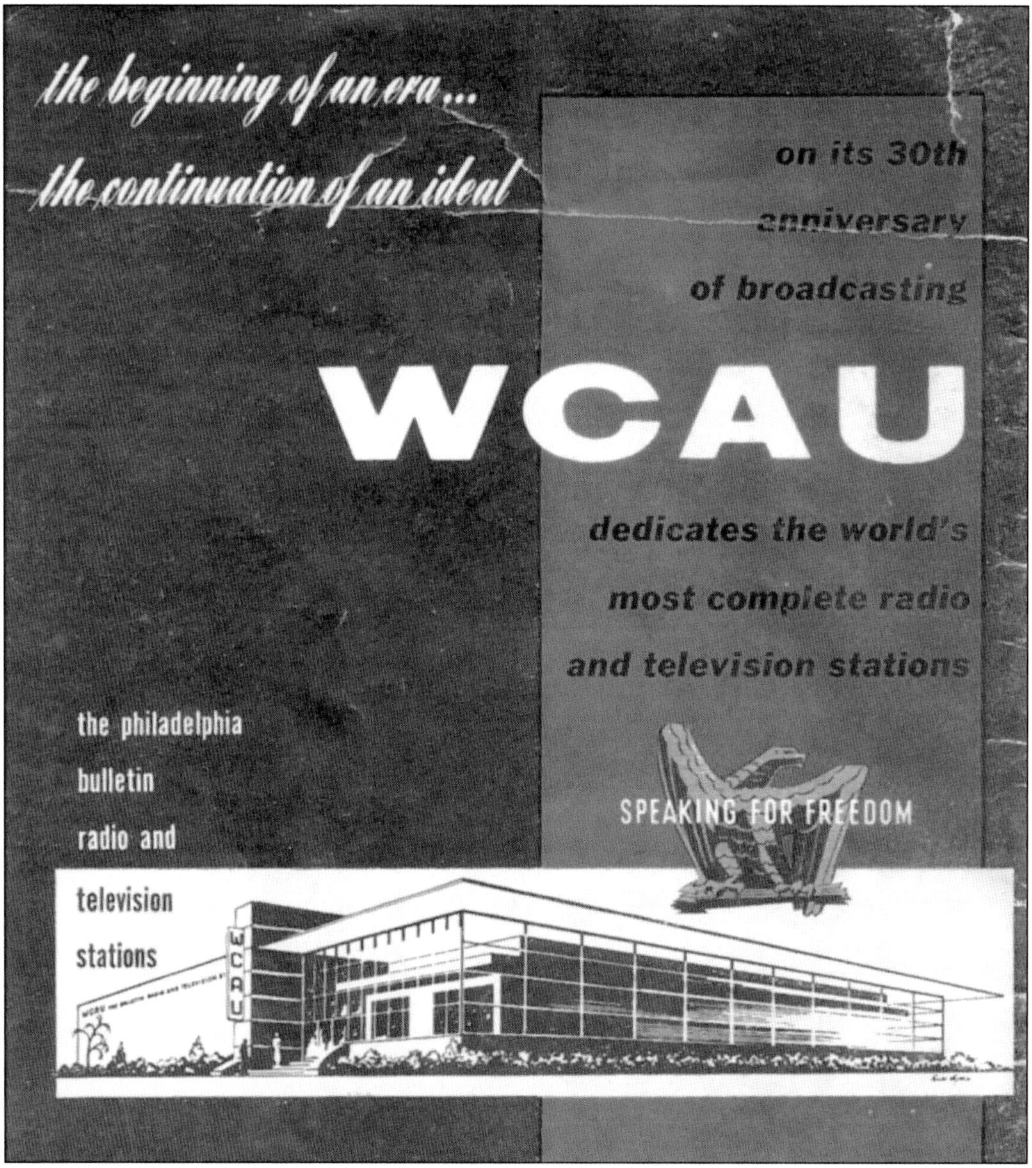

TV Digest ran this promo for WCAU's new building in May 1952. The station is scheduled to move to Center City in 2017, leaving the future of this historical building in doubt. (WCAU-TV.)

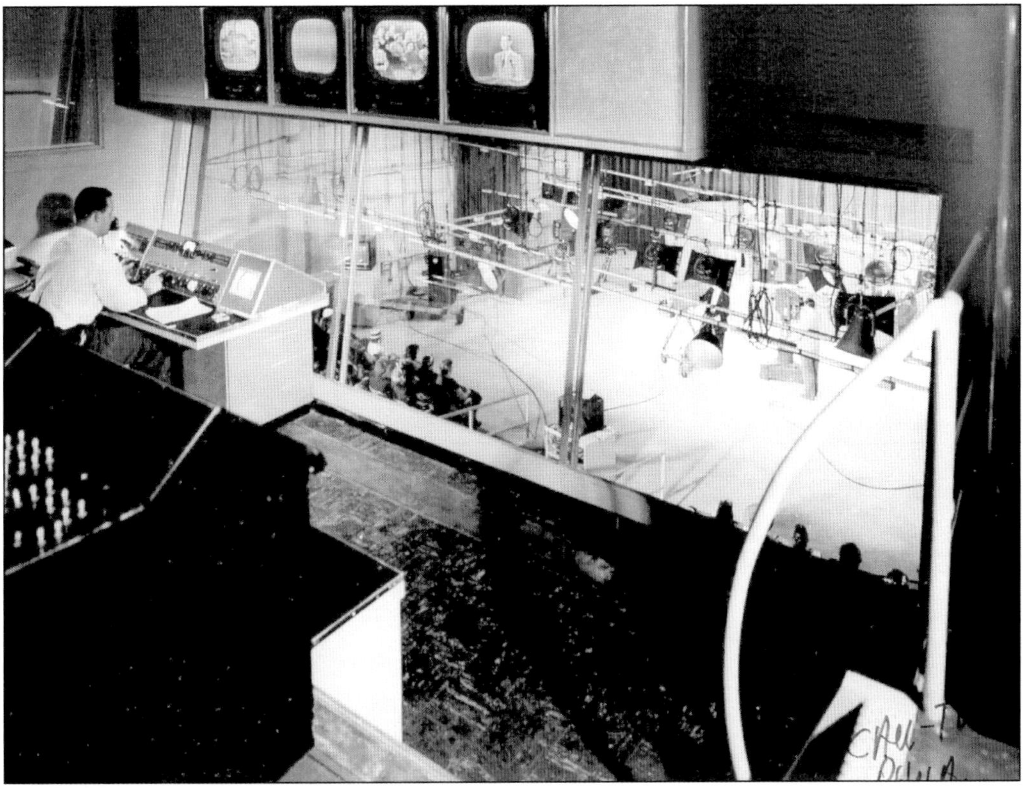

One of the control rooms in the early days of WCAU's City Avenue building is pictured here. The director's console is in the upper left. The audio panel is in the left foreground. Most recently, the old control rooms have been used for storage or offices. (WCAU-TV.)

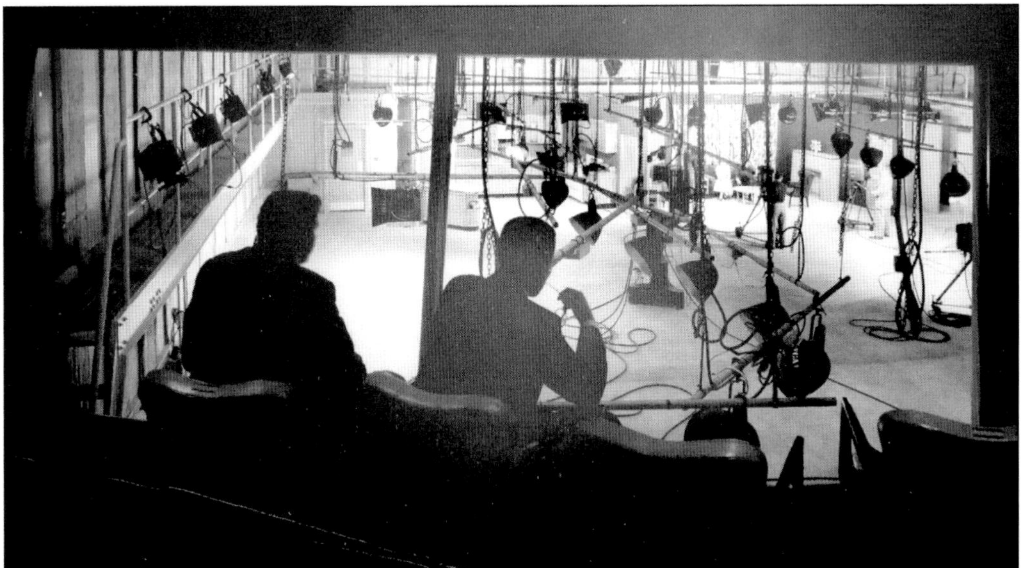

Here is another view of a studio looking down from a control room at WCAU. Two of the three original studios are now used for the newsroom, satellite center, and editing suites. (WCAU-TV.)

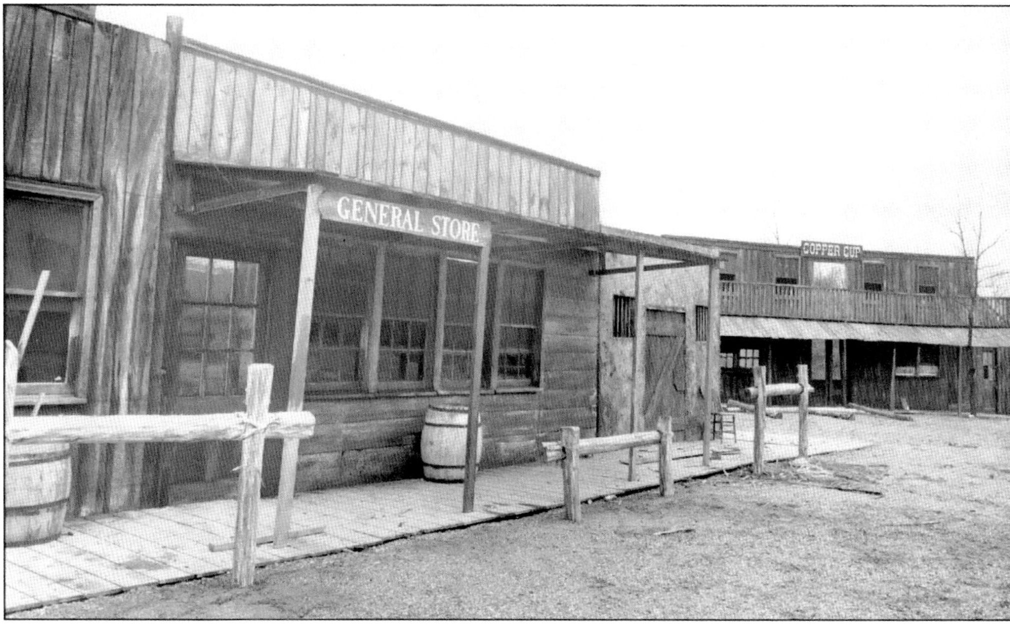

The fictional Western town of Huberle, Montana, population 486, was the home of *Action in the Afternoon*. Huberle was on the back lot of WCAU. The show was network TV's first Western and first program to air live on CBS stations across the country. This area is now a parking lot. (WCAU-TV/Special Collections Research Center, Temple University Libraries, Philadelphia, Pennsylvania.)

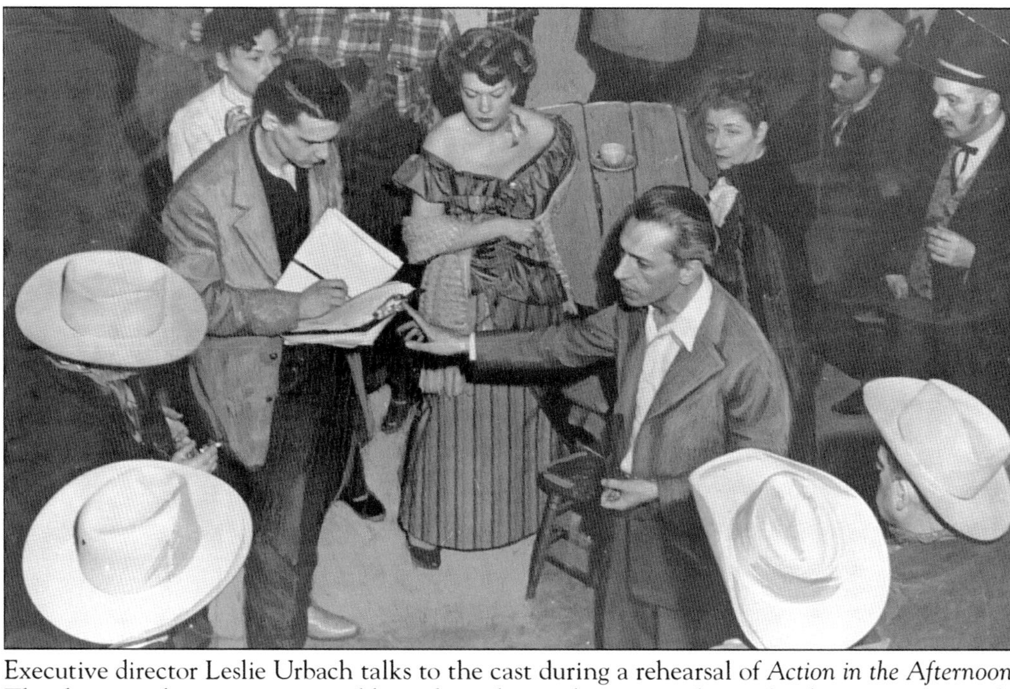

Executive director Leslie Urbach talks to the cast during a rehearsal of *Action in the Afternoon*. The show was live, so actors would run from the outdoor set to the studio during commercials. *Action in the Afternoon* aired from February 1953 to January 1954. (WCAU-TV/Special Collections Research Center, Temple University Libraries, Philadelphia, Pennsylvania.)

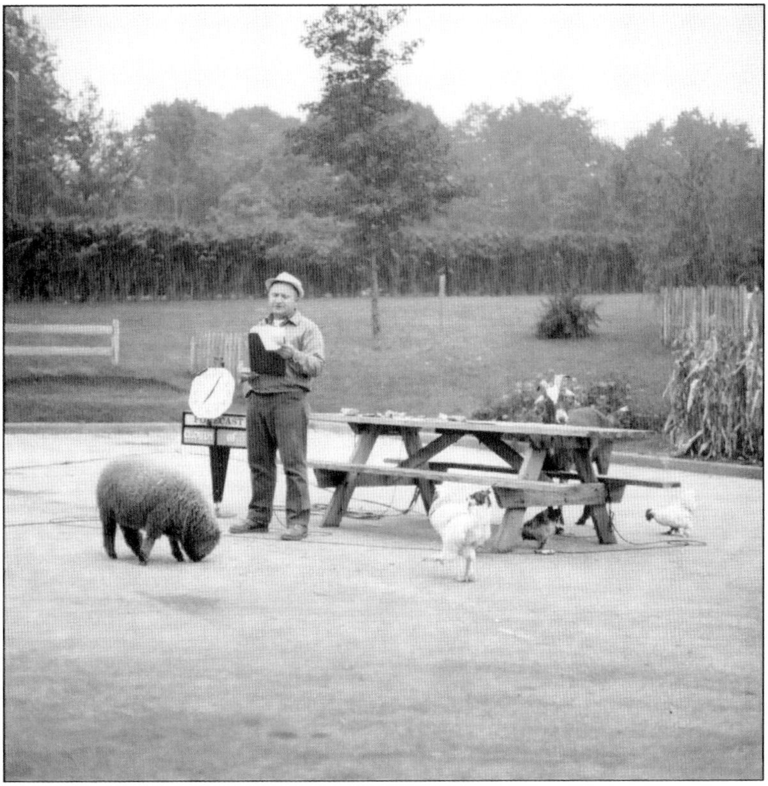

Another show broadcast from WCAU's back lot was the *Bill Bennett Farm Show*. This picture shows Bill Bennett on his set with some friends. The show originally aired at 6:00 a.m., then at 6:45 a.m., during its run from 1960 to 1965. The cornfield in the background is now an office complex. (WCAU-TV/Special Collections Research Center, Temple University Libraries, Philadelphia, Pennsylvania.)

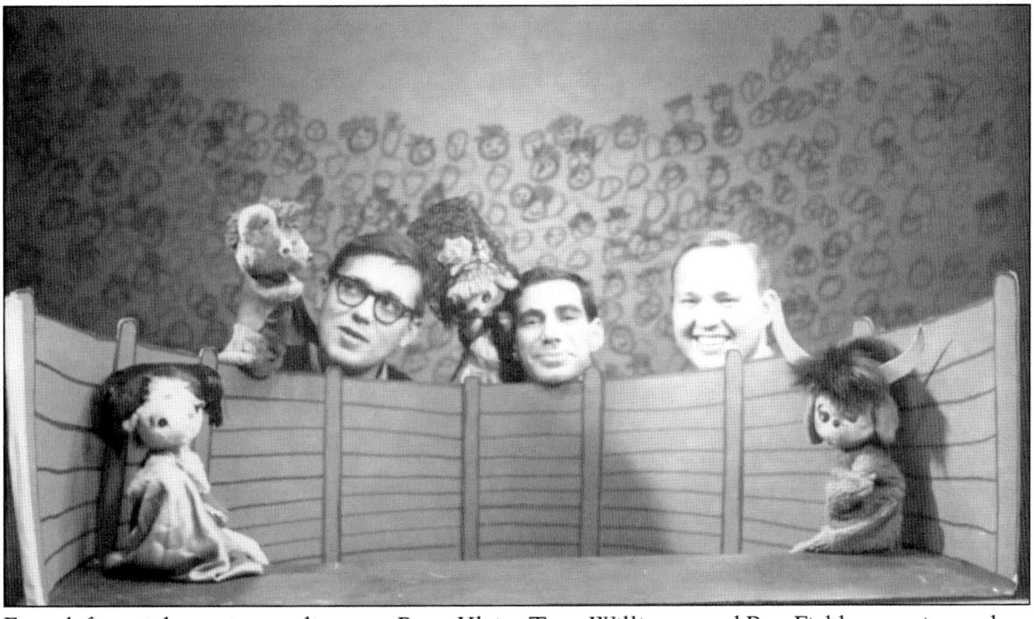

From left to right, assistant directors Peter Klein, Tony Williams, and Ray Fielder are pictured on the set of *Tottle* around 1964. *Tottle* was a taped, 30-minute children's puppet show. Jane Norman, better known as Pixanne, and Marshall Izen were the voices and puppeteers. *Tottle* was one of only two local programs that aired on WCAU in the days following the assassination of Pres. John F. Kennedy. The episode talked about the death of loved ones. (WCAU-TV.)

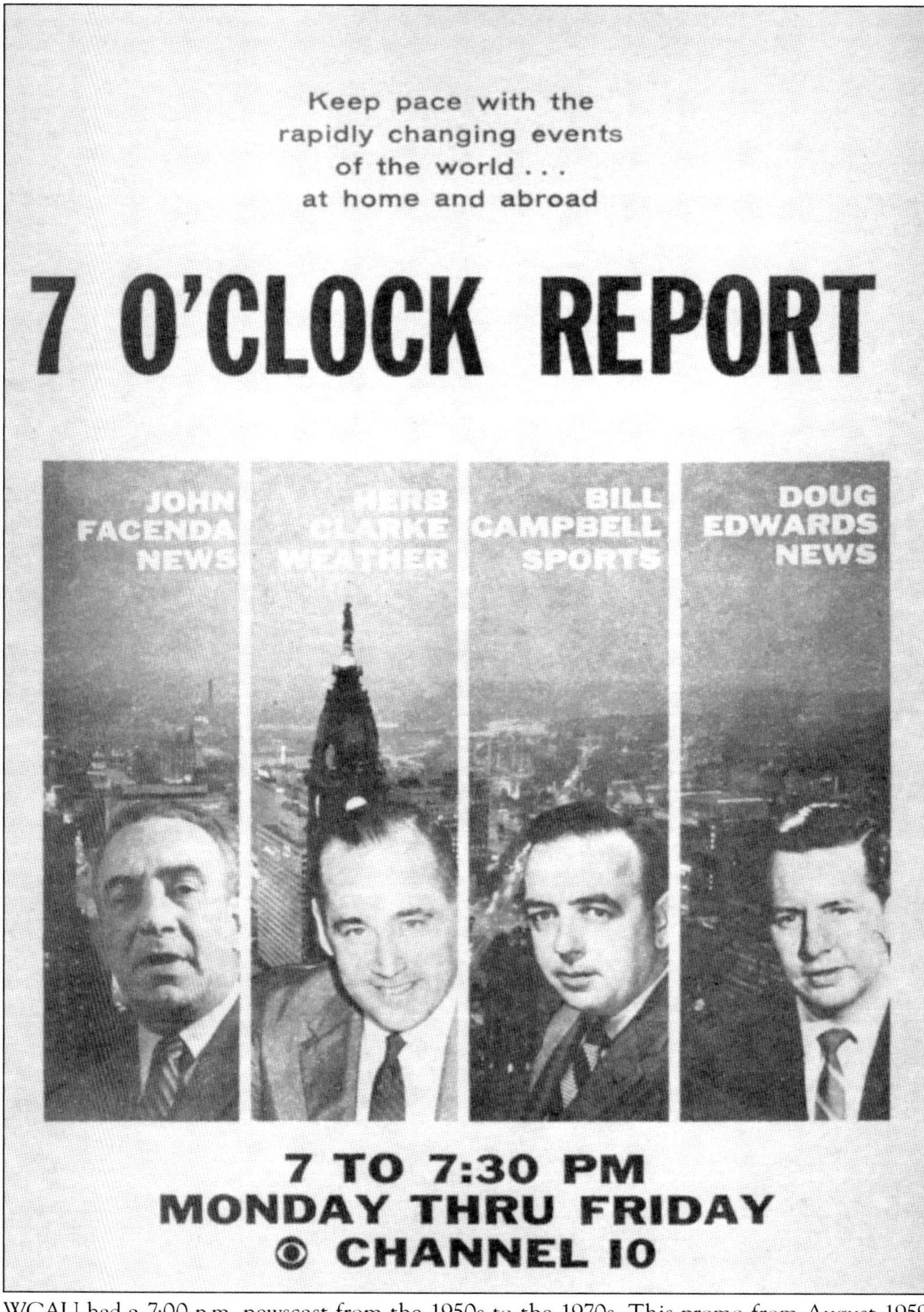

WCAU had a 7:00 p.m. newscast from the 1950s to the 1970s. This promo from August 1959 shows, from left to right, anchorman John Facenda, weatherman Herb Clarke, and sportscaster Bill Campbell. Doug Edwards anchored the *CBS Evening News* in between the local newscasts. (WCAU-TV.)

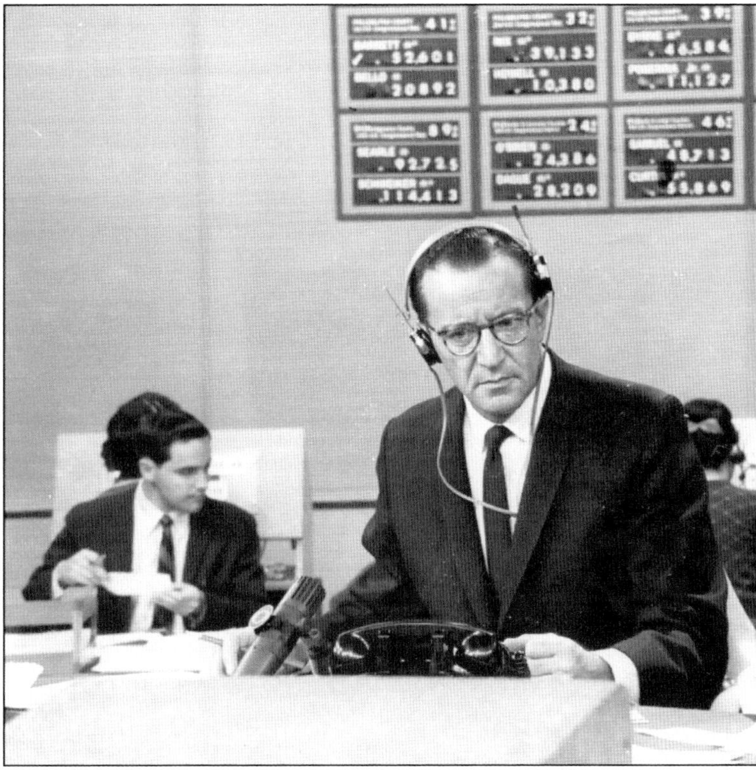

Gene Crane anchors Channel 10's election night coverage in 1964. Crane transitioned from host to news anchor and reporter in the 1960s. He also served as a weatherman on the station's noon newscasts. (WCAU-TV.)

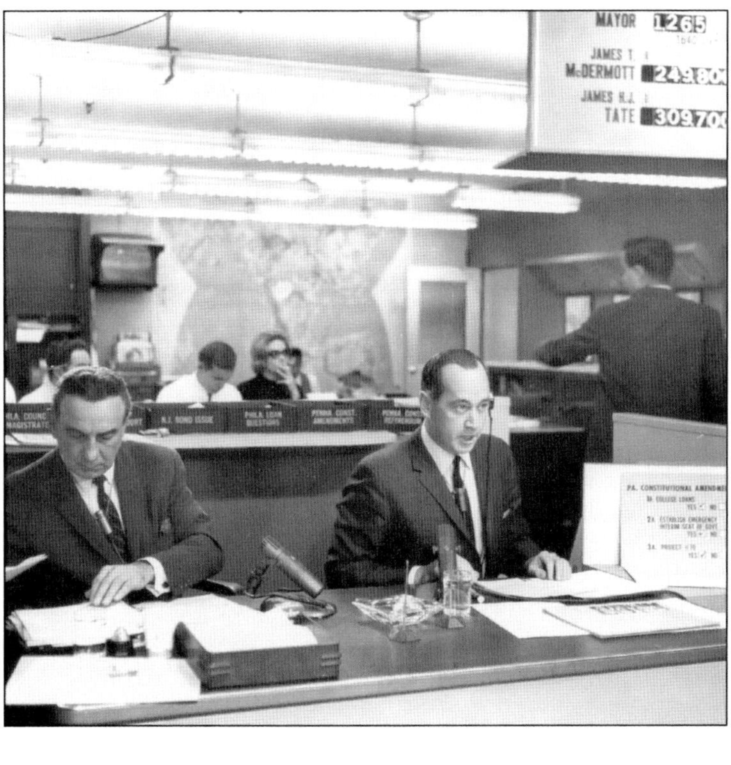

Anchors John Facenda (left) and Herb Clarke deliver election night returns on November 5, 1963. Clarke was also WCAU's lead weatherman. The board on the upper right shows results for the Philadelphia mayoral race, the big contest of the night. (WCAU-TV.)

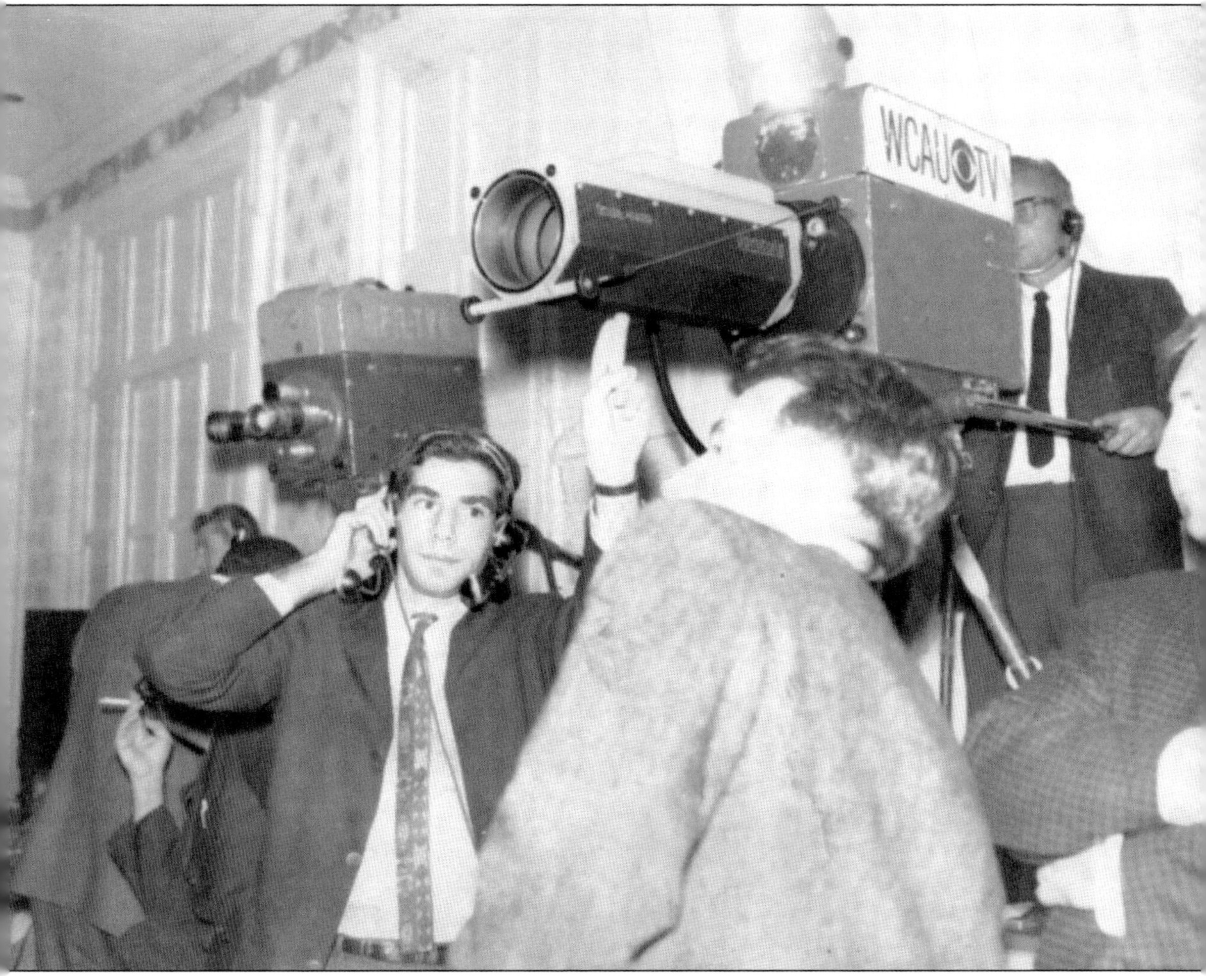

This is another view of WCAU's election night coverage in 1963. Stage manager Tony Williams (wearing a headset) directs the station's remote broadcast from the Bellevue-Stratford Hotel in Center City. Note the large studio camera used for the broadcast. (WCAU-TV.)

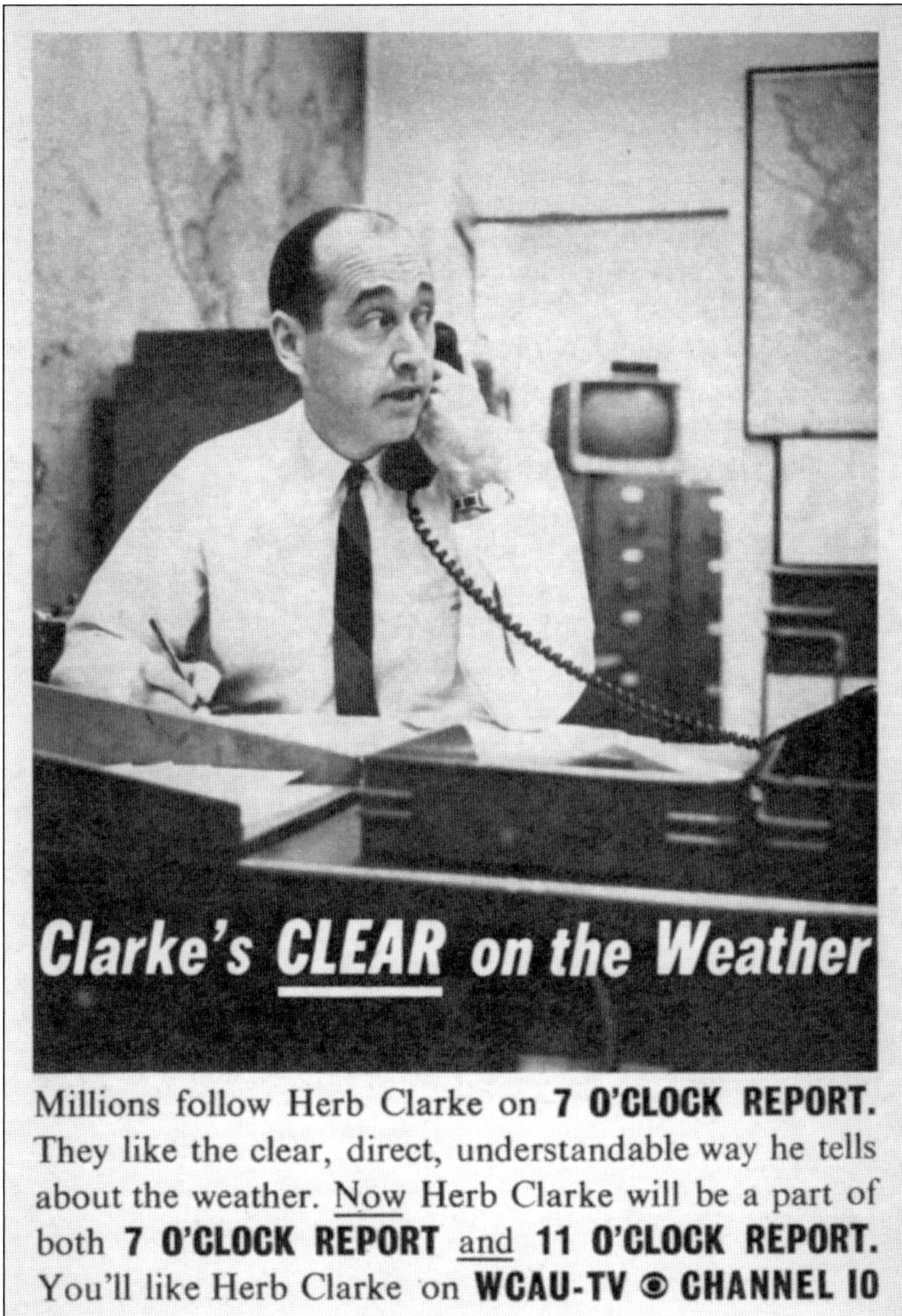

Herb Clarke was WCAU's weatherman from 1958 to 1997. He was known as the "Atlantic Weatherman," because he wore an Atlantic Gas Station uniform while doing the weather in the late 1950s. Clarke passed away in 2012. (WCAU-TV.)

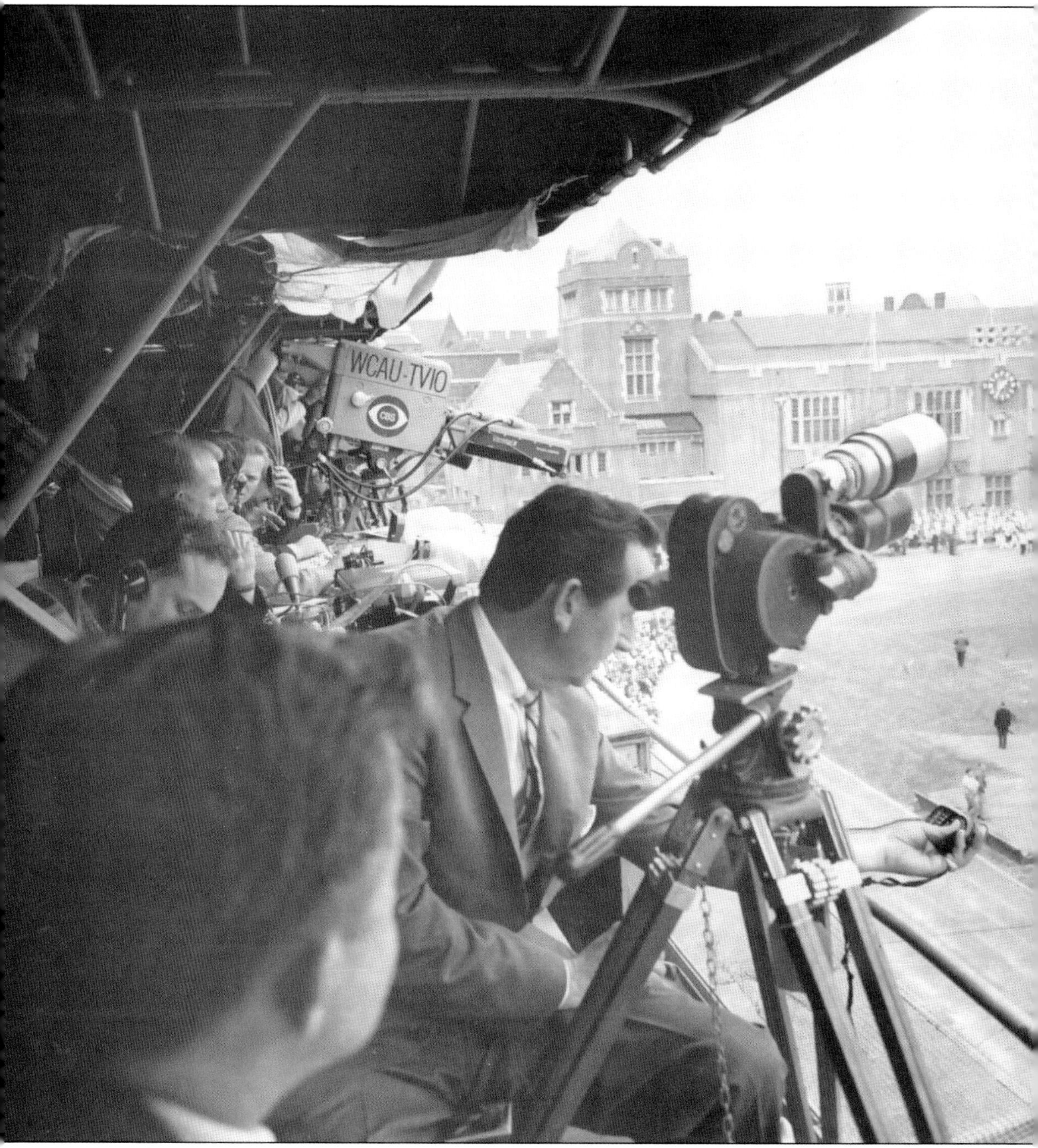

WCAU was a pioneer in local sports coverage. This image shows a WCAU camera during a broadcast of the Eagles and 49ers game on October 11, 1964, at Franklin Field. CBS Network sportscasters Jack Whitaker and Tom Brookshier both began at Channel 10 and both did play-by-play for the Eagles. (WCAU-TV/Special Collections Research Center, Temple University Libraries, Philadelphia, Pennsylvania.)

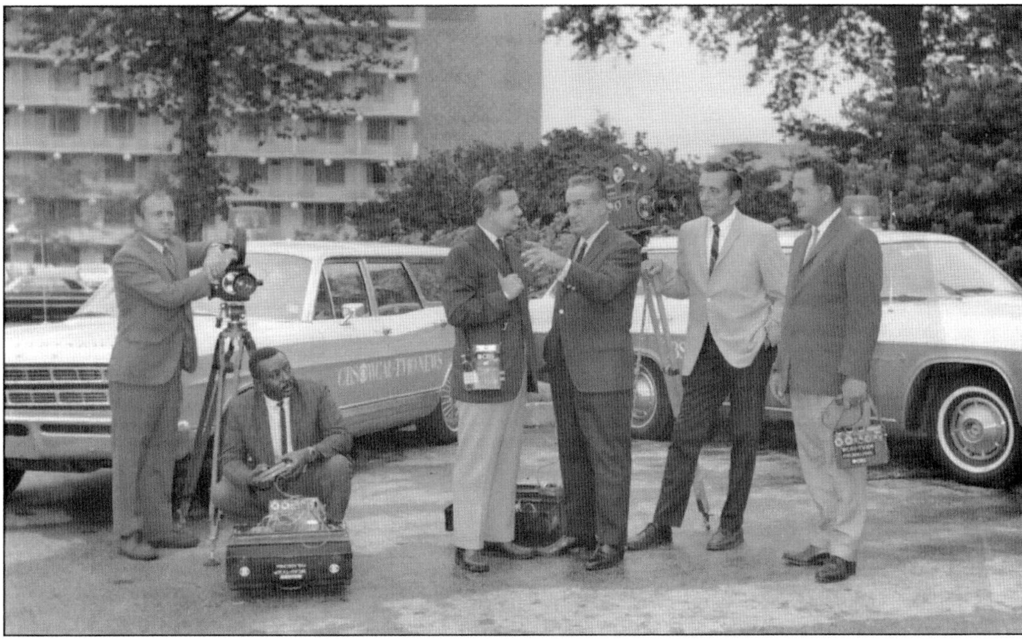

The Channel 10 news team, anchored by John Facenda (pointing), poses in front of the station's news cars in the mid-1960s. This publicity photograph was taken in the back parking lot of the station's City Avenue building. (WCAU-TV.)

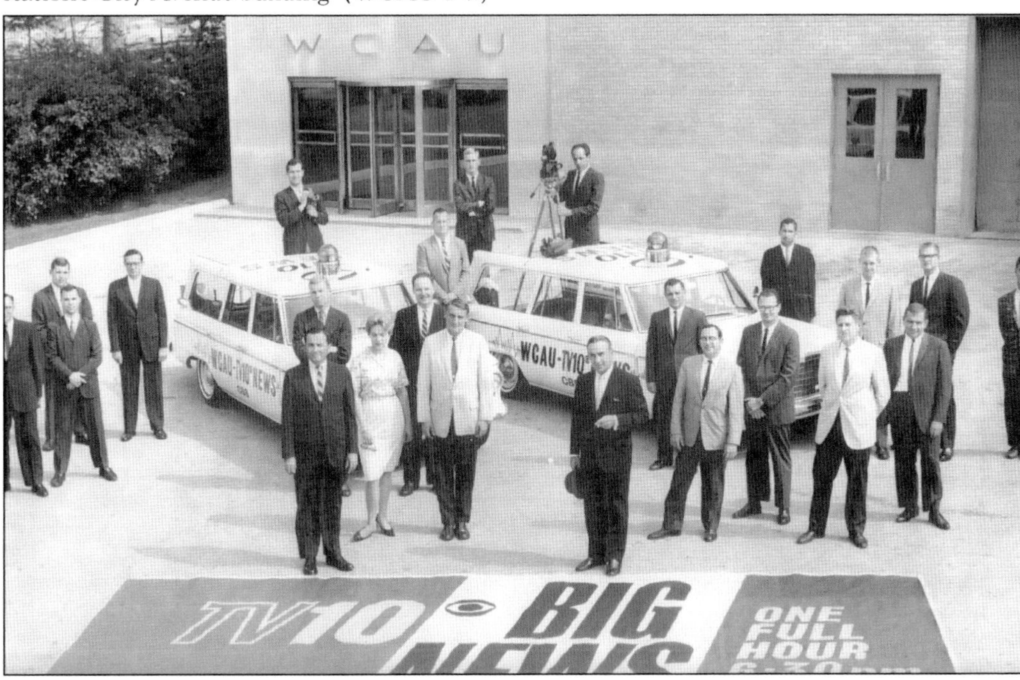

This is another team photograph of Channel 10's reporters and photographers with John Facenda at center, holding hat. Notice the film cameras and tape recorders, state of the art equipment for the station's camera crews. The door under the WCAU sign leads to a long ramp that goes into the front lobby. Most recently, the door was used as the audience entrance for the *10! Show.* (WCAU-TV.)

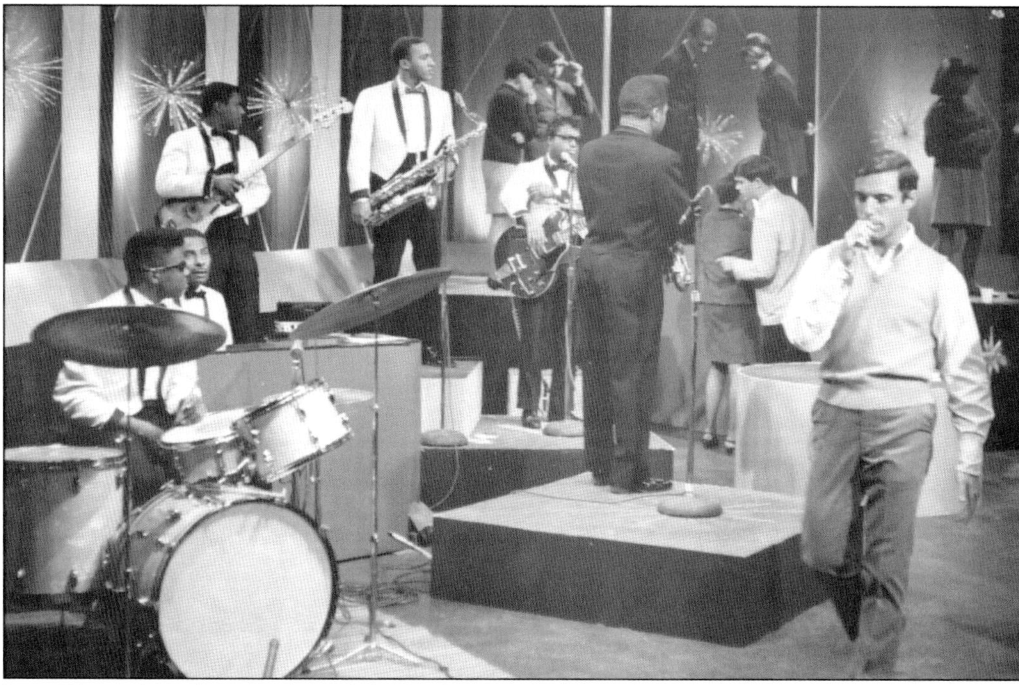

Jerry Blavat (standing on the right), also known as "the Geator with the Heater" and "the Boss with the Host Sauce," hosted *The Discophonic Scene* from 1965 to 1966. The show was produced by WCAU and syndicated nationally. Blavat always got the top musical guests of the time to appear, including James Brown. (WCAU-TV.)

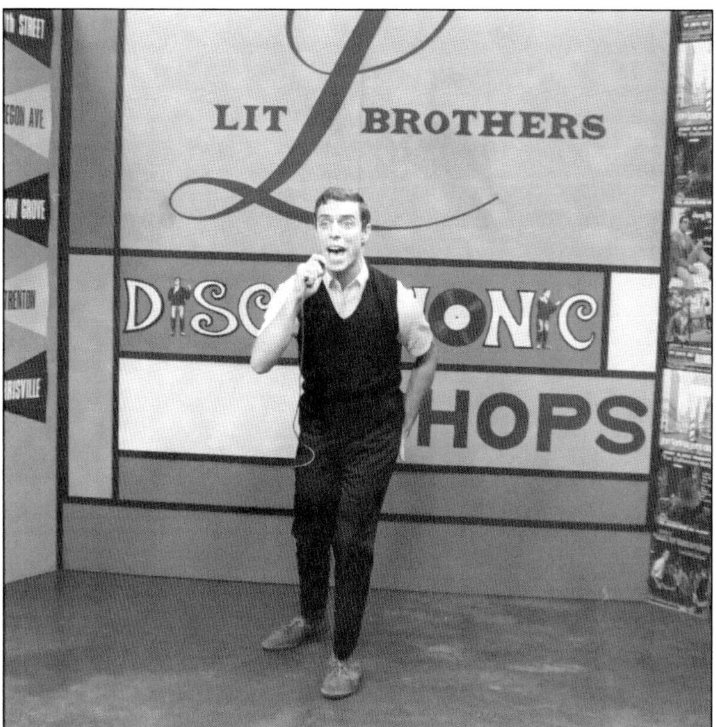

The Discophonic Scene aired live every Saturday afternoon from the Lit Brothers Department Store on Sixty-ninth Street in Upper Darby. Blavat started as a dancer on WFIL's *Bandstand* in the early 1950s. (WCAU-TV.)

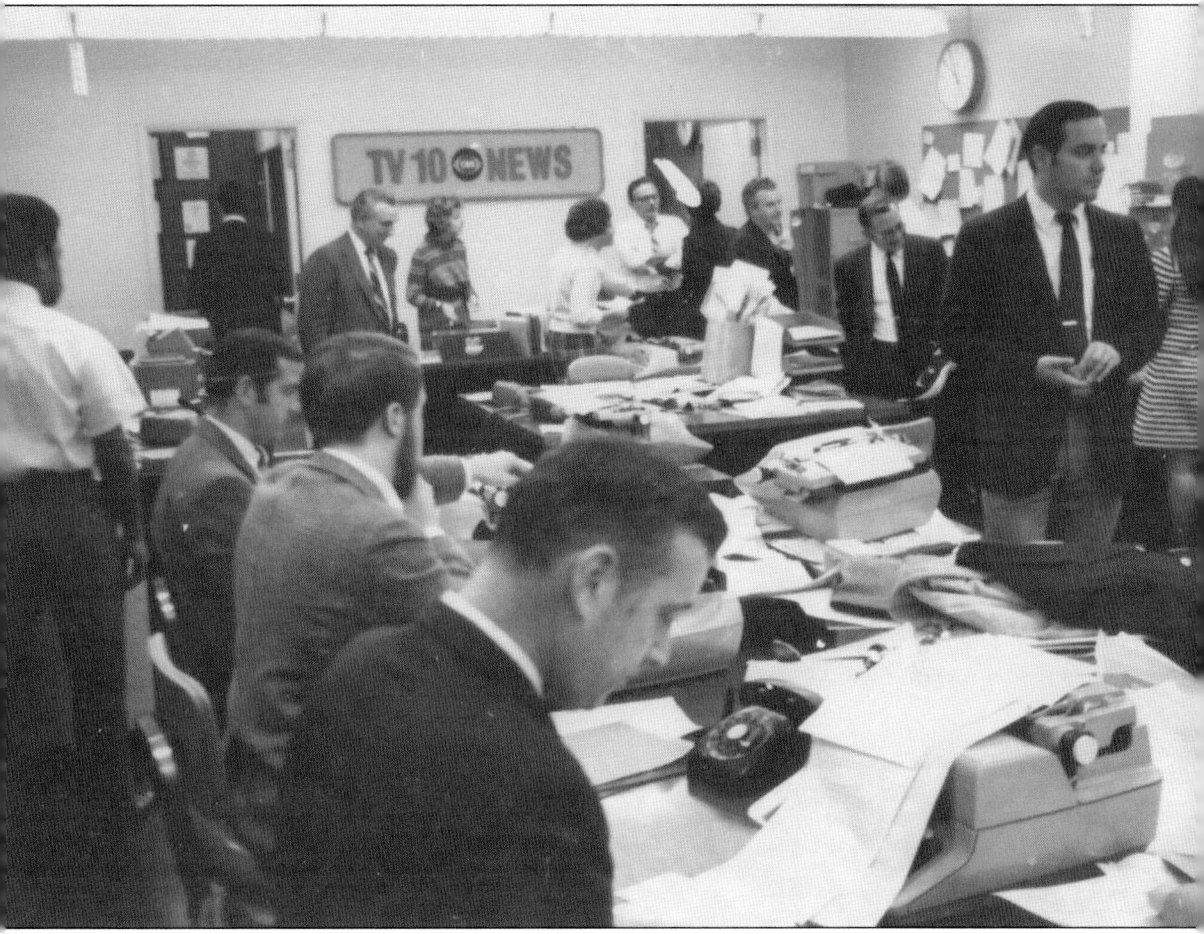

This photograph from the late 1960s shows WCAU journalists at work in the newsroom. In the foreground is Herb Clarke; two seats to his left is reporter Bill Baldini, and John Facenda stands under the "TV" sign in the background. News director Barry Nemcoff stands with his arms crossed at the back of the room. (WCAU-TV.)

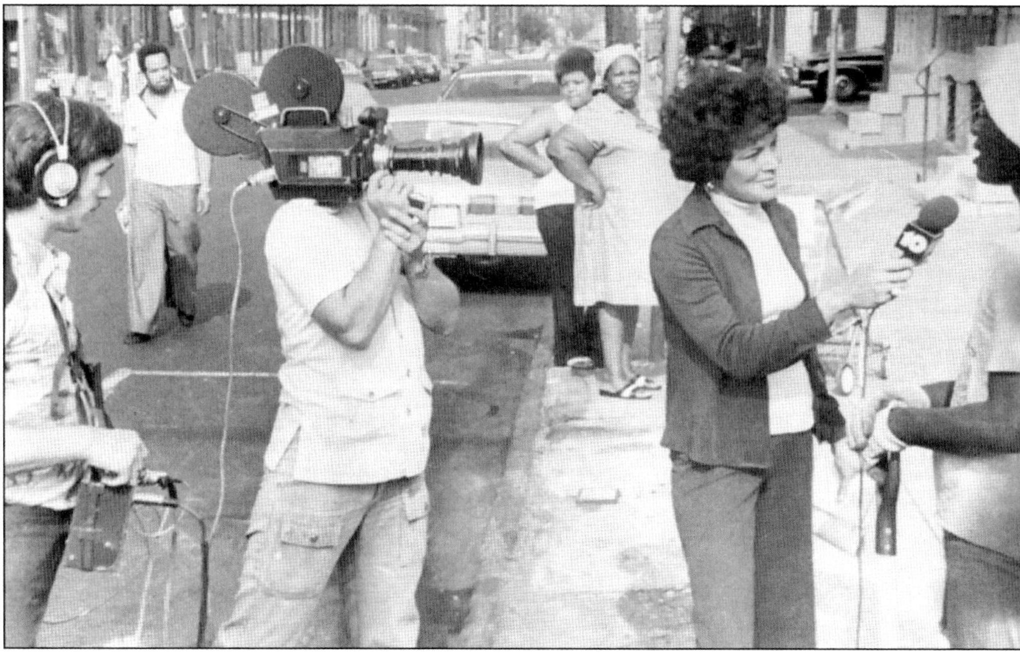

Legendary WCAU reporter Edie Huggins interviews someone for a news story in the late 1970s. "Miss Edie," as she was known to her coworkers, was one of the first African American women on Philadelphia television. She worked at Channel 10 from 1966 until her death in 2008. In the early 1970s, Huggins hosted her own talk show called *Morning Side*. (WCAU-TV.)

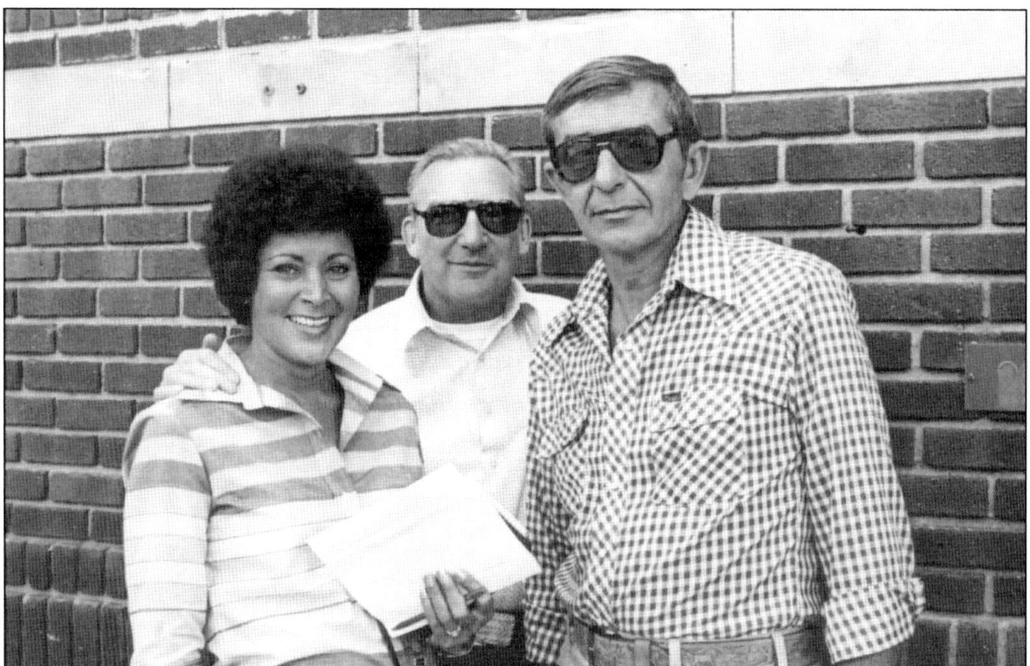

Huggins poses with WCAU technician George Schumaker (center) and photographer Joe Mayer on July 26, 1977, at the Wissahikon Center in Philadelphia. WCAU proclaimed March 30, 2006, Edie Huggins Day to honor her 40 years at the station. (WCAU-TV.)

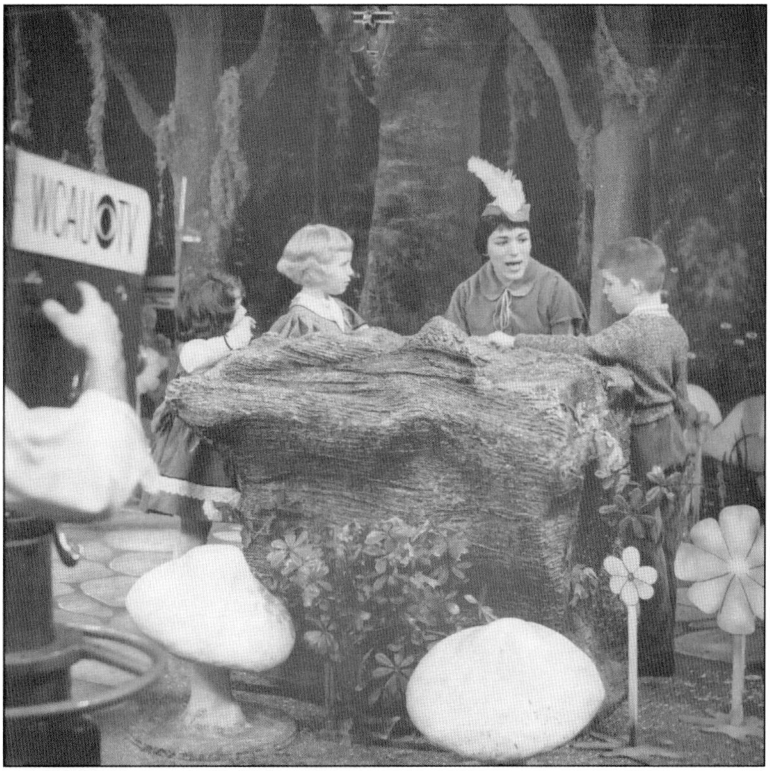

Pixanne was one of the most popular children's shows in Philadelphia TV history. Pixanne was played by Jane Norman, seen here holding court in the enchanted forest. Norman opened every showing "flying" into the studio on cables; she says they only dropped her once. (WCAU-TV.)

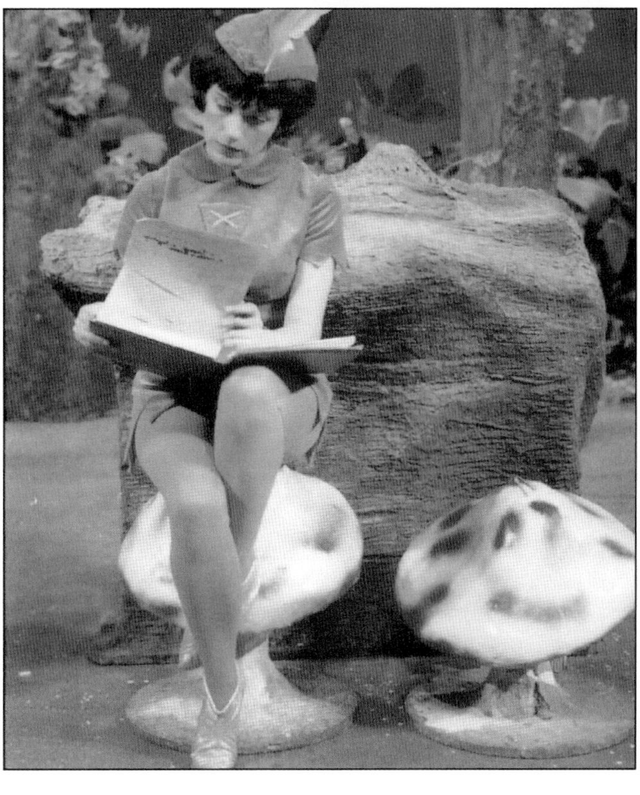

Jane Norman goes over the show rundown for another episode of *Pixanne*. She first pitched her concept for the show to WFIL-TV but was rejected. So, Norman walked across the street to WCAU. She did not have an appointment but asked to talk to the program director. Channel 10 loved the idea and hired Norman. *Pixanne* aired on WCAU from 1960 to 1969 and another seven years after that in syndication. (WCAU-TV.)

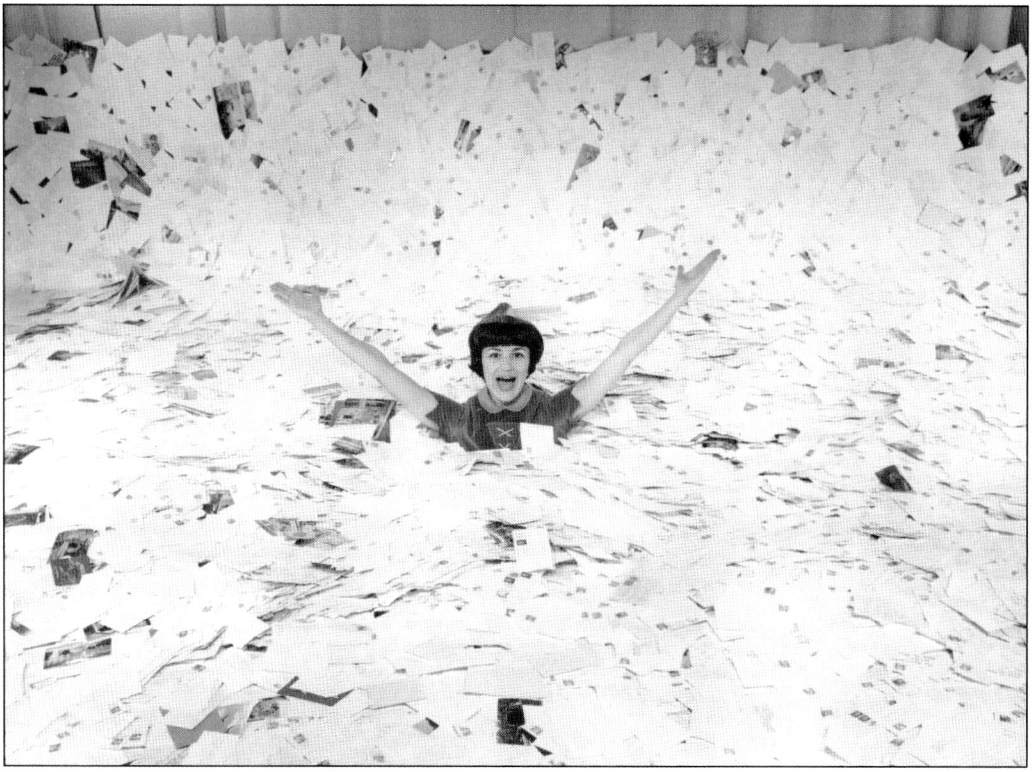

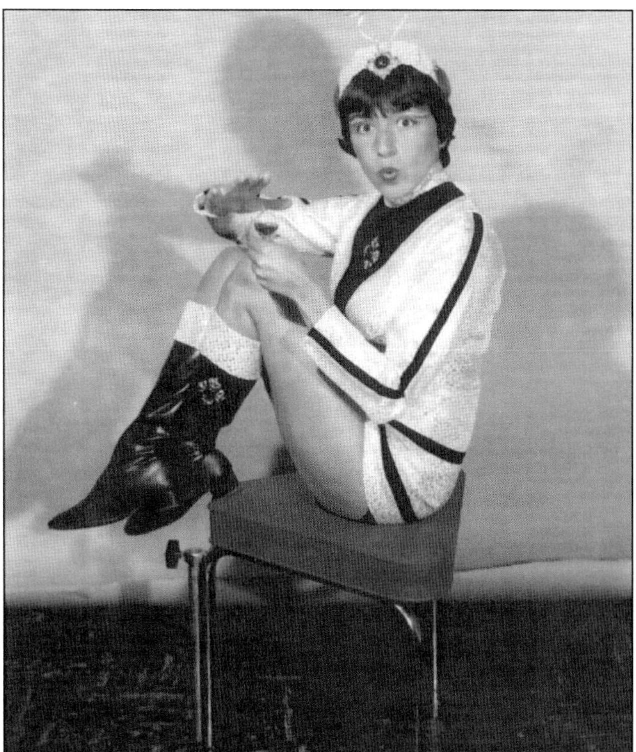

Here is proof that *Pixanne* had a lot of fans. Jane Norman is surrounded by more than 20,000 entries she received for a contest. Norman is also an accomplished musician and singer. (WCAU-TV.)

Some viewers may not remember that Jane Norman had another show besides *Pixanne*. She transformed into Space Girl for a Saturday morning sci-fi show of the same name in the 1960s. (WCAU-TV.)

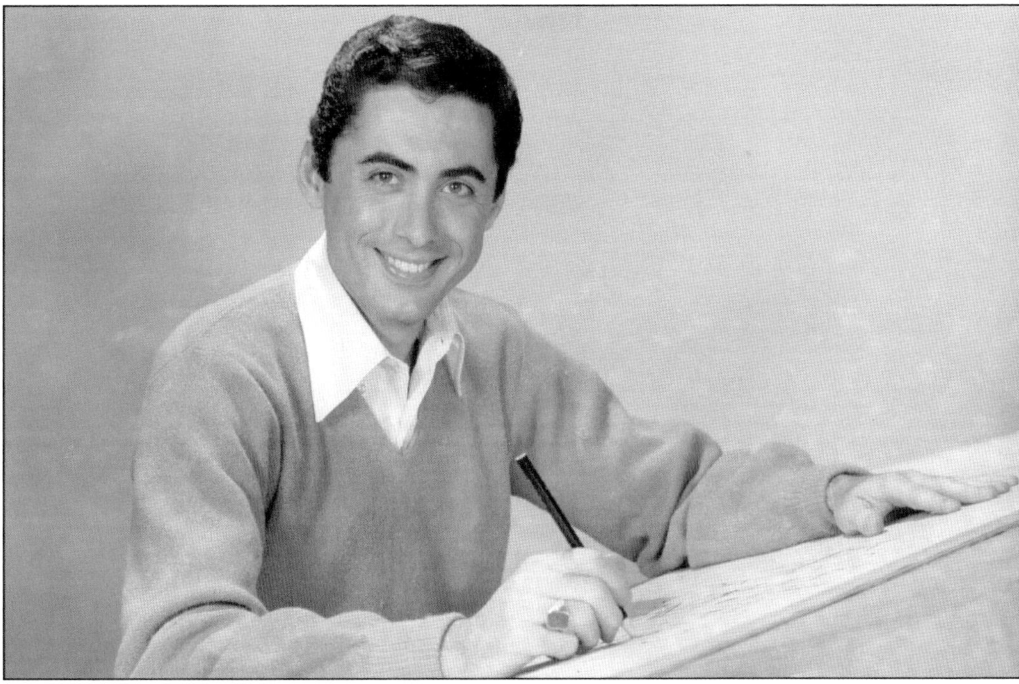

Another staple of mornings on Channel 10 was Gene London. London hosted several programs, the most famous being *Cartoon Corners*. He was an accomplished artist and storyteller. London used a large drawing pad to illustrate characters to the audience of children who sat around him. (The Gene London Collection.)

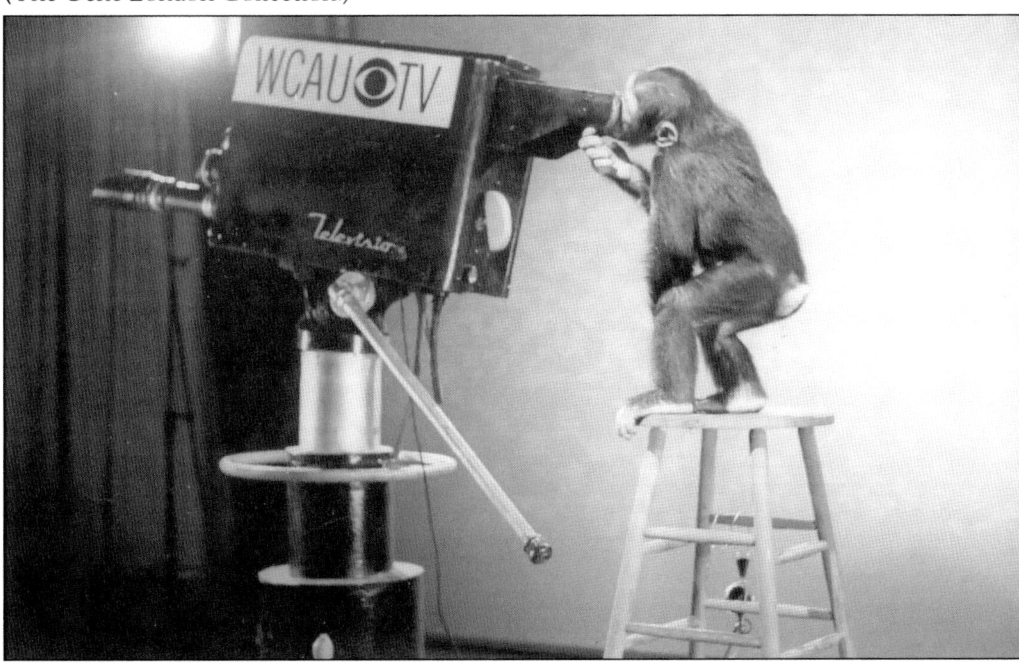

One of Gene London's guests gets up close and personal with a WCAU studio camera. London hosted children's shows on Channel 10 from 1959 to 1977. He is now a successful fashion designer. (WCAU-TV.)

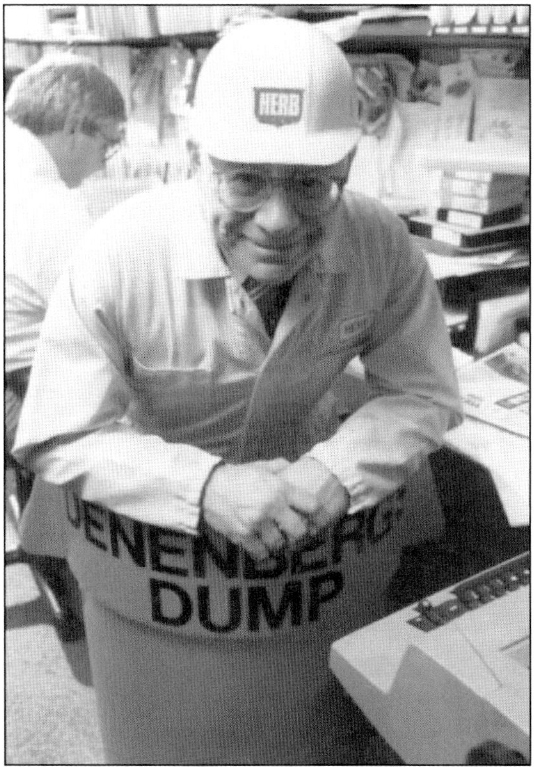

Herb Denenberg was a former Pennsylvania state insurance commissioner who made a huge impact on Philadelphia TV with his "Denenberg's Dump" segments on WCAU. He spent almost 20 years fighting for consumers and testing products. The bad ones would go into the "dump." In this photograph, behind Denenberg is his producer Ed Dress, who spent 31 years at WCAU. (WCAU-TV.)

Pictured is a typical WCAU news crew in the late 1970s. From left to right are cameraman Joe Mayer (with a film camera), reporter Joan Dinerstein, and soundman Ralph Rodeo. Dinerstein was WCAU's main female anchor in the late 1970s. Mayer's daughter says her father made his makeshift camera brace from an old office chair. (WCAU-TV.)

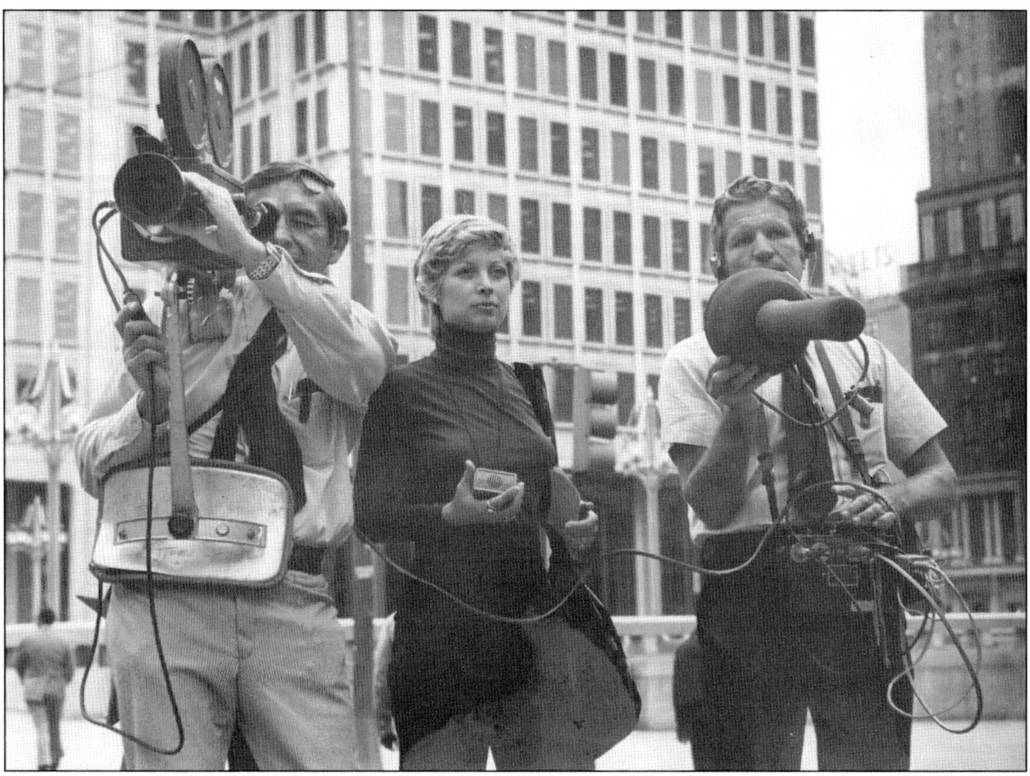

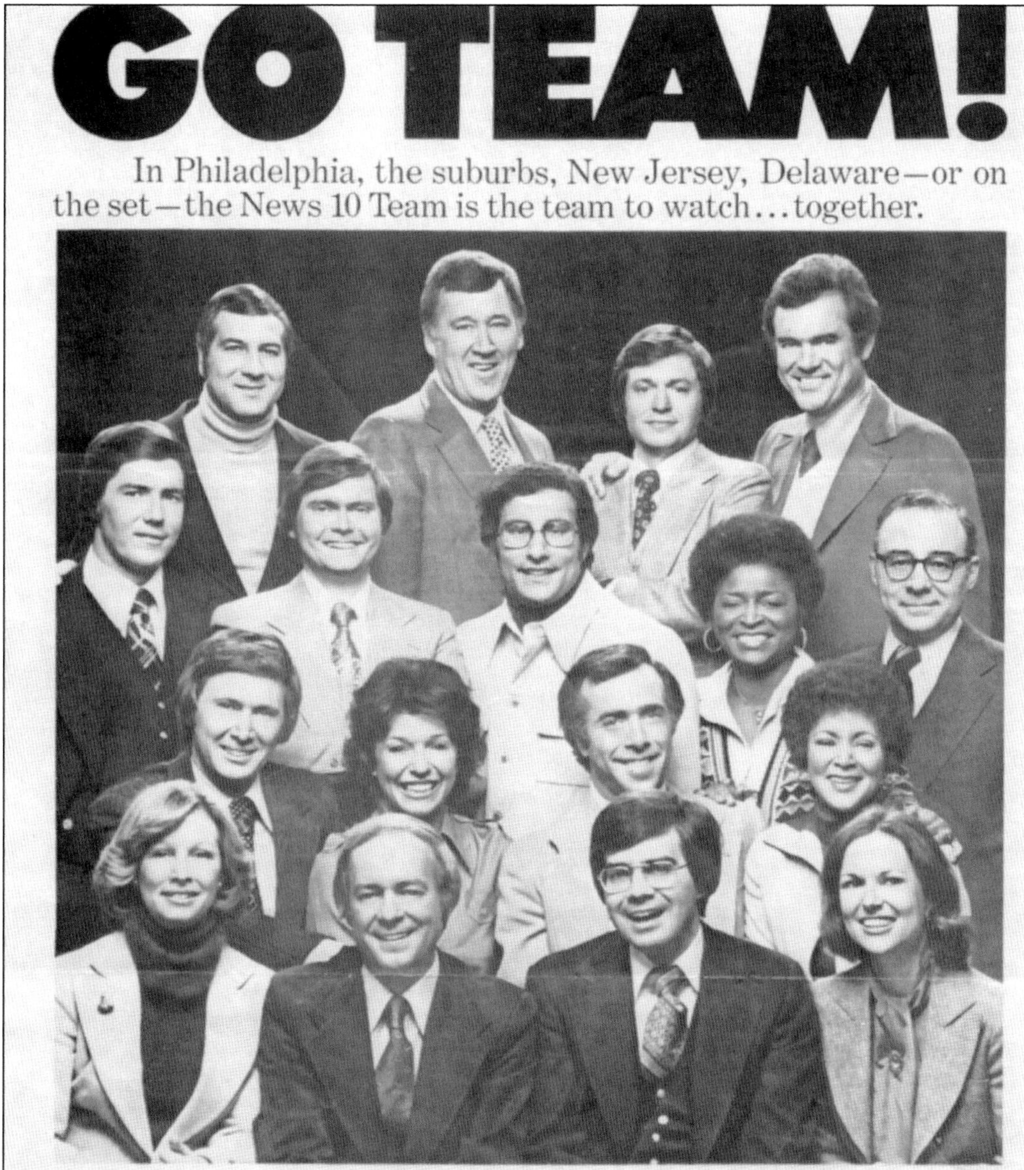

This print ad shows the News 10 team from the late 1970s. Ralph Penza and Joan Dinerstein were the main anchors until Larry Kane arrived from his brief stint at ABC in New York. (WCAU-TV.)

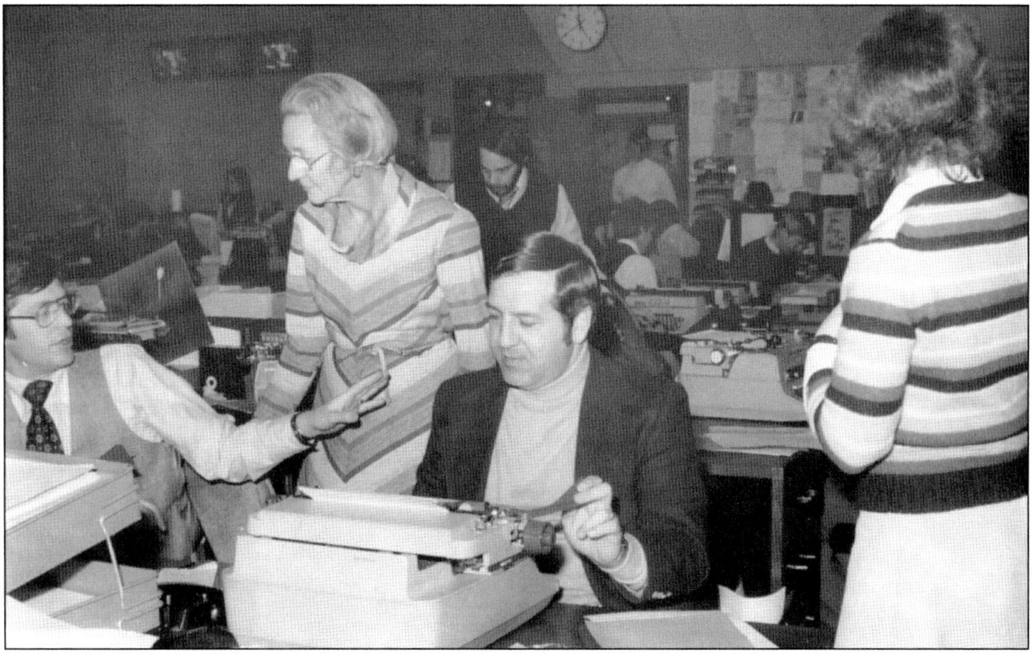

Terry Ruggles (left) and Bill Baldini work at their typewriters in the WCAU newsroom in the late 1970s. Standing behind them (wearing the vest) is Gary Herman, a producer at the station from 1978 to 2014. All three men were responsible for making Channel 10 a leader in local news coverage. (WCAU-TV.)

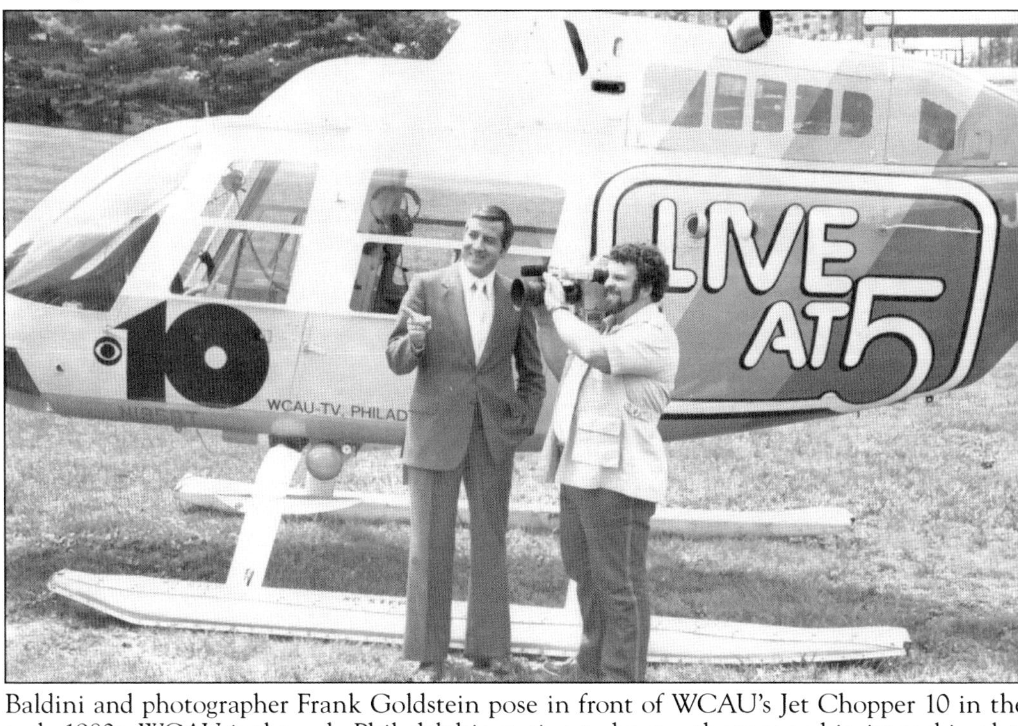

Baldini and photographer Frank Goldstein pose in front of WCAU's Jet Chopper 10 in the early 1980s. WCAU is the only Philadelphia station to have a chopper pad in its parking lot. (WCAU-TV.)

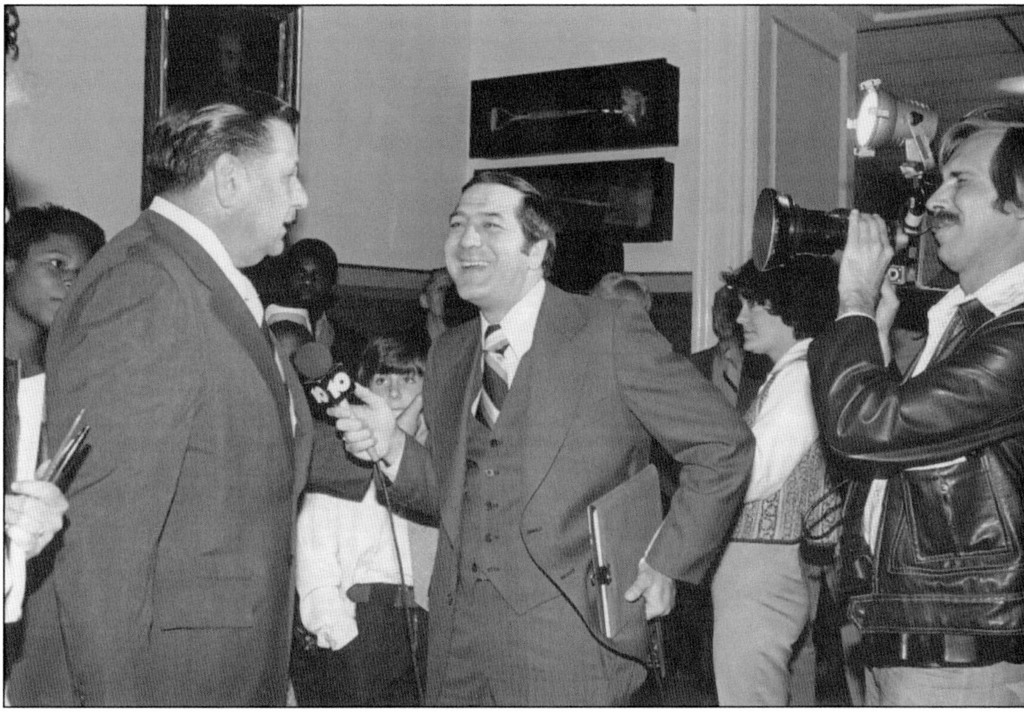

Baldini interviewed almost every newsmaker in the Delaware Valley in his 40 years at WCAU. He says one of his favorites was Philadelphia mayor Frank Rizzo. This photograph was taken outside the mayor's office in Philadelphia City Hall. (WCAU-TV.)

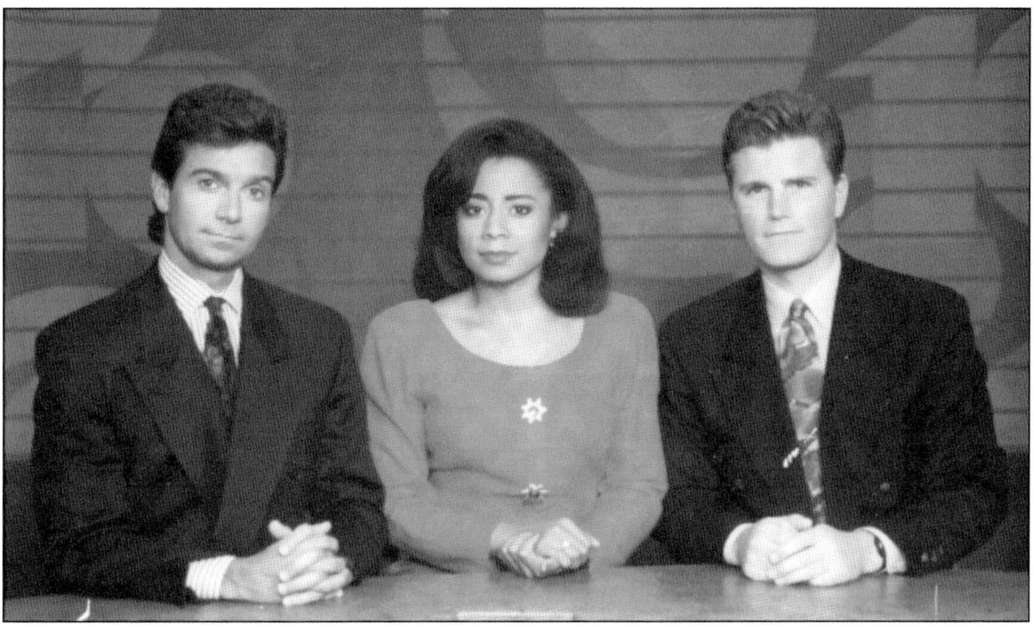

Meteorologist John Bolaris (left) and anchors Renee Chenault-Fattah (center) and Tim Lake (right), along with sports director Vai Sikahema (not pictured) were the faces of *NBC 10 News* starting in 2003. Lake took over as the main male anchor when Larry Mendte bolted for KYW. (WCAU-TV.)

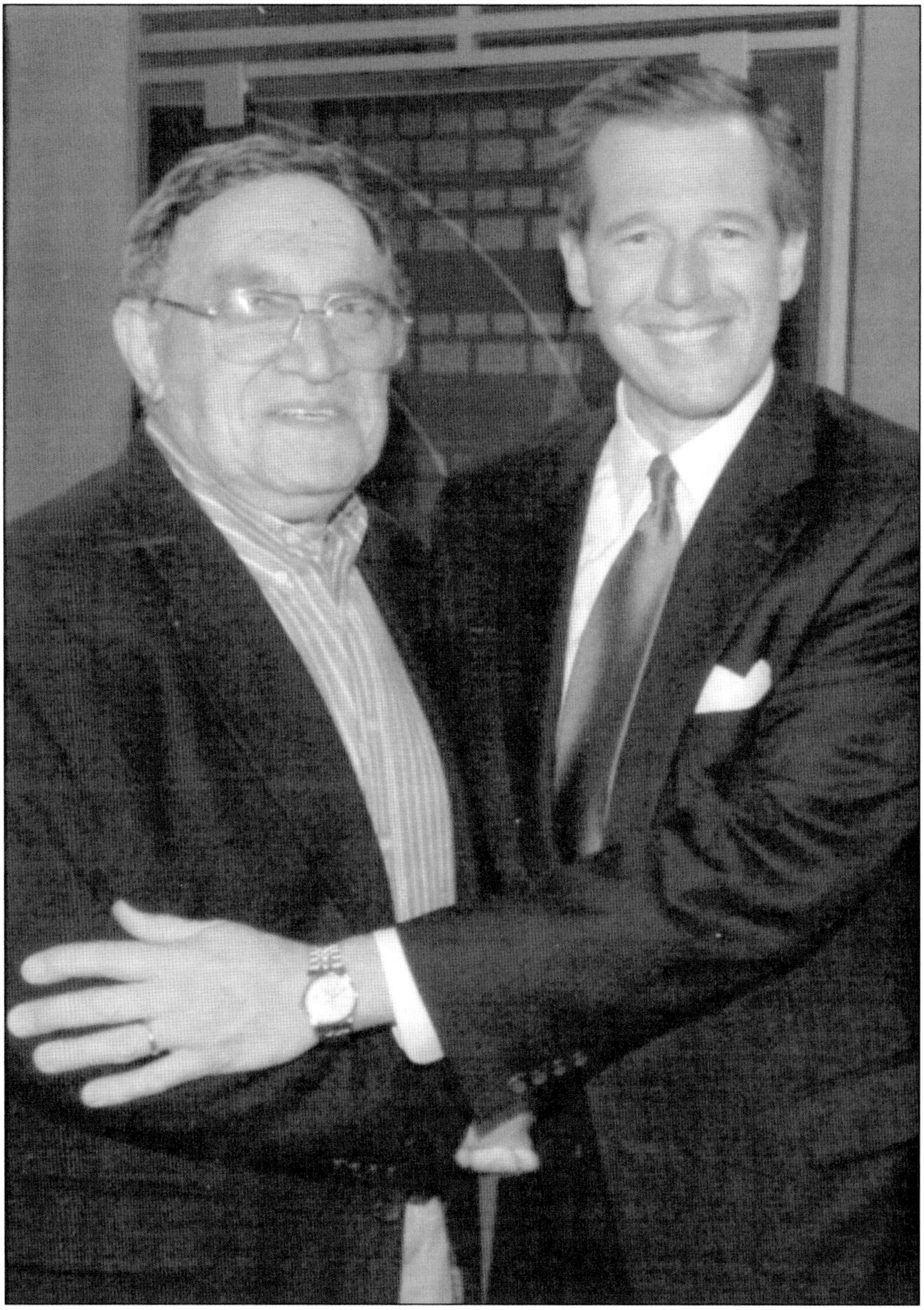

NBC Nightly News anchor Brian Williams covered New Jersey for WCAU in the mid-1980s. The man on the left is Dave Gomberg (known to everyone as Dave Neal). Gomberg was an assignment editor and producer for KYW, WPVI, WCAU, and WTXF (possibly the only person to have ever worked for all four news operations). Gomberg helped shape the careers of Williams, Larry Kane, Jessica Savitch, and many, many others. He came up some of Philadelphia TV's most popular news features: "The Sidewalk Gourmet" with Hank Sperka at Channel 6, "Denenberg's Dump" at WCAU, and the "Energy Warden" with KYW's Dick Sheeran. (WCAU-TV.)

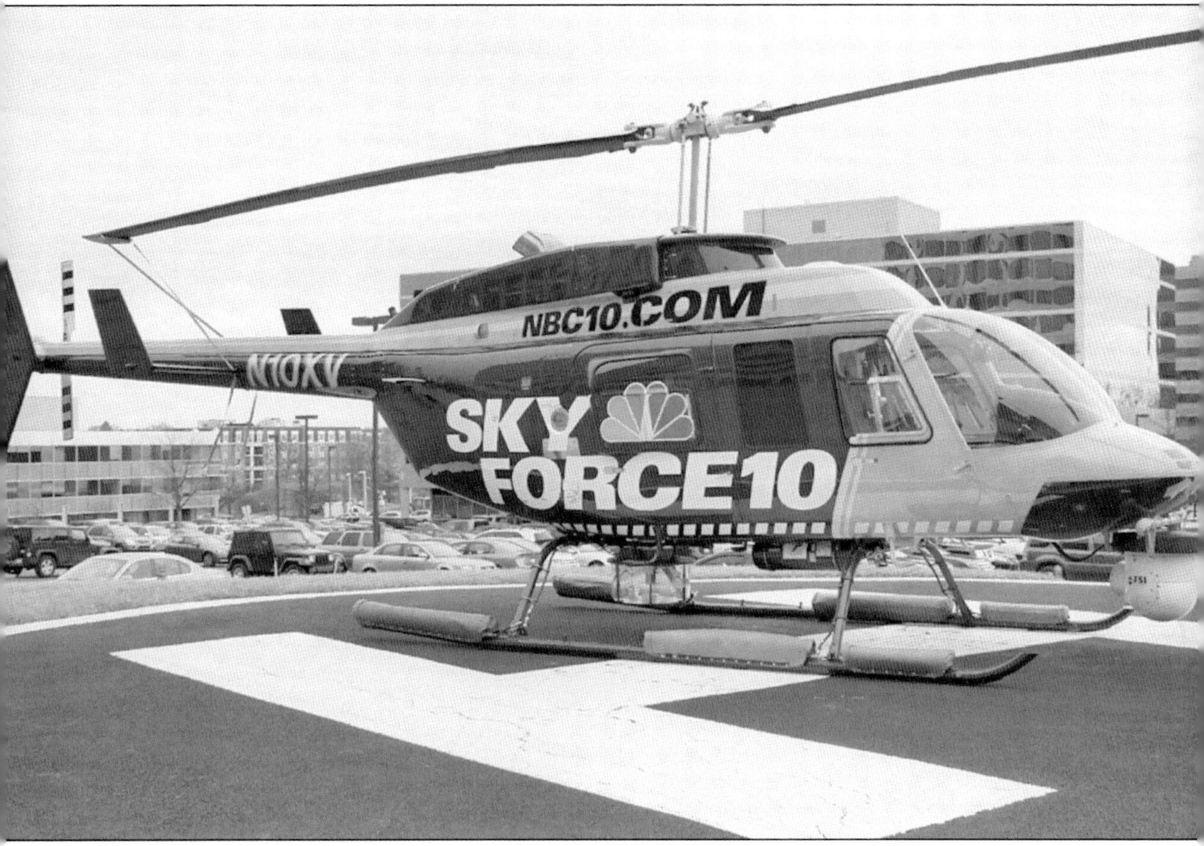

Skyforce 10 is pictured parked on the landing pad behind WCAU-TV. The station is the only one in the city with a landing area at its building. The helicopter has five cameras inside and out and captures video in high definition. It went into service in 2013. (Author's collection.)

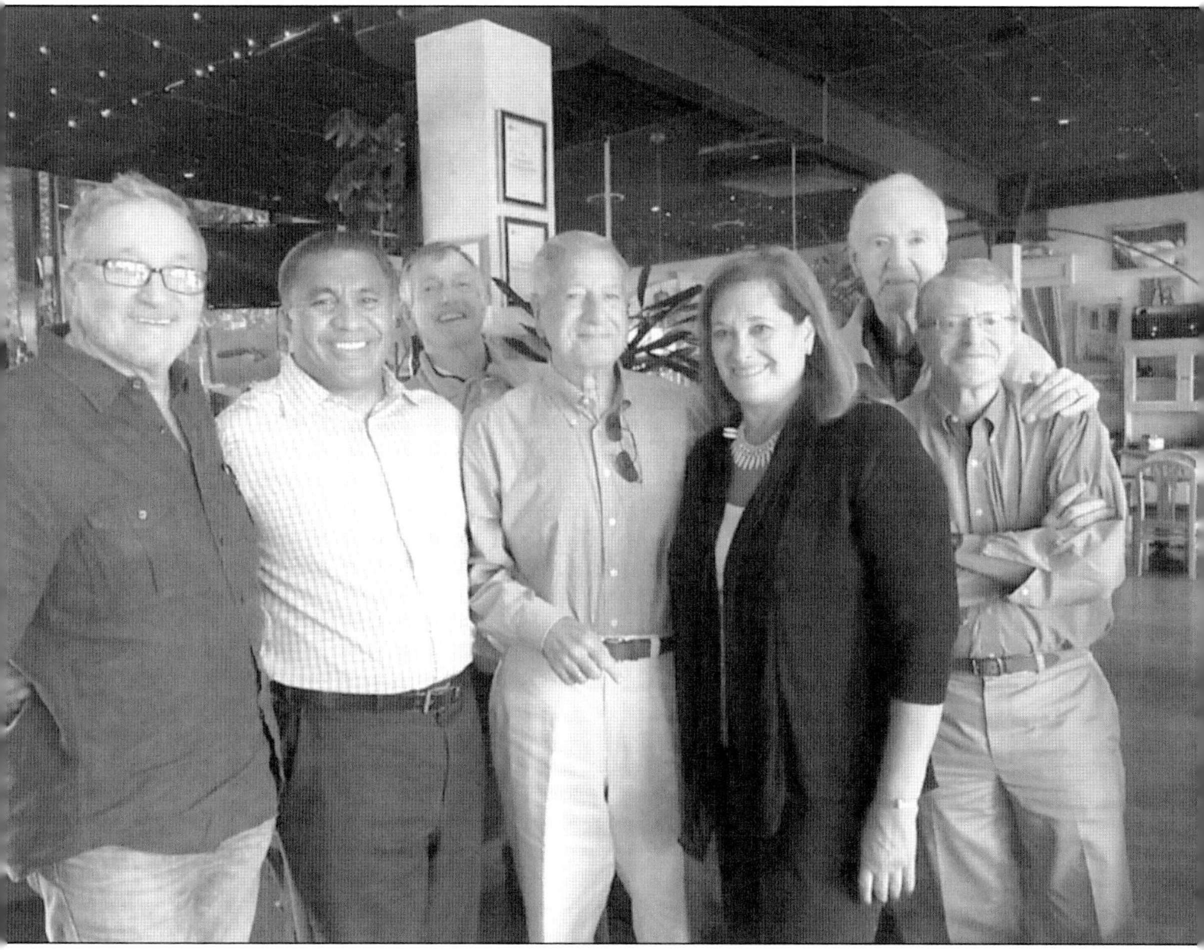

The Mount Rushmore of WCAU includes, from left to right, former reporter Mike Strug, NBC 10 morning news anchor and former Philadelphia Eagles player Vai Sikahema, former morning news anchor Steve Levy, former reporter Bill Baldini, former medical reporter Cherie Bank, former sportscaster Big Al Meltzer, and former morning news anchor and reporter Terry Ruggles. This picture was taken in September 2014. (Cherie Bank.)

Four

WHYY Channel 35/12

WHYY began broadcasting educational programs into Philadelphia school classrooms. It was first on Channel 35 at a time when few TV sets were able to pick up the UHF band. In 1957, WHYY moved down the dial to Channel 12. This was made possible thanks to an FCC deal with the owners of channel 12 in Wilmington, Delaware. The station's first headquarters was the old WCAU building on Chestnut Street. The station moved to the former WFIL building at Forty-sixth and Market Streets in 1969. The facility was donated by Channel 6 and *Philadelphia Inqurier* owner Walter Annenberg. WHYY moved to its current home, the former Living History Center Museum on Independence Mall, in 1979. WHYY is licensed to Wilmington-Philadelphia. It maintains a studio in Delaware's largest city as well. WHYY also once produced a nightly Delaware news show from the Wilmington studios. The station is the regional Public Broadcasting Service (PBS) station for the city and all of Delaware. Local legend Bill Webber ended his long career hosting several documentaries on Channel 12, including *A Walk Up Broad Street*. Ed Cunningham and Patrick Stoner have each hosted many shows on WHYY and are often seen on the station's annual pledge drives asking for viewer donations. WTXF meteorologist Sue Serio and her husband, former WTXF sportscaster Bill Vargas, were both part of the TV 12 news team in the 1990s, when the station broadcast a nightly 5:30 newscast. In 2000, WHYY became one of the nation's first PBS stations to begin digital broadcasting.

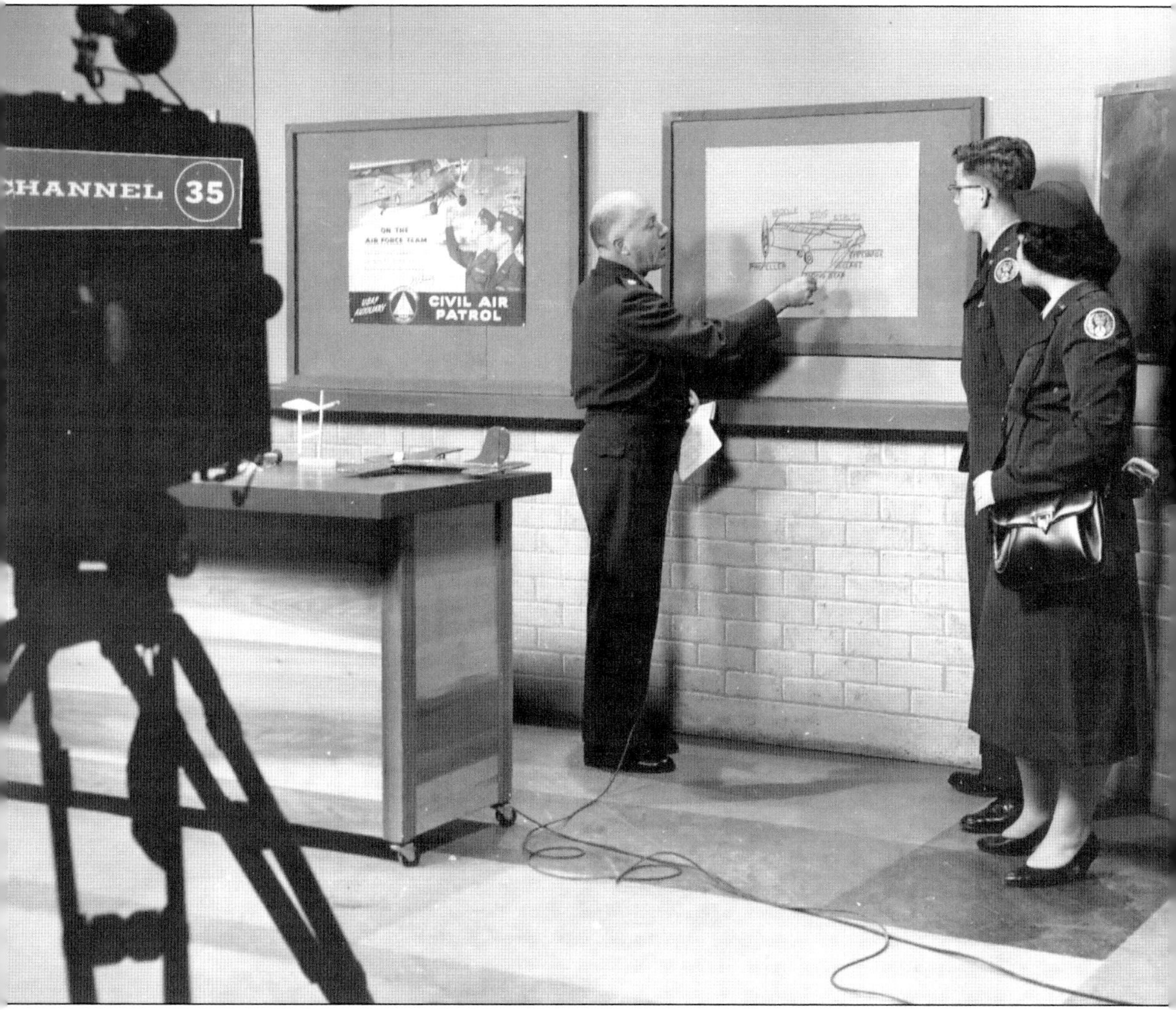

WHYY signed on September 16, 1957, on Channel 35. The station originally broadcast programs into Philadelphia School District classrooms from 9:00 a.m. to 3:00 p.m. This picture shows Major Herr from the Civil Air Patrol during a segment of the *Zach Bernstein Show*, an educational science program. (WHYY.)

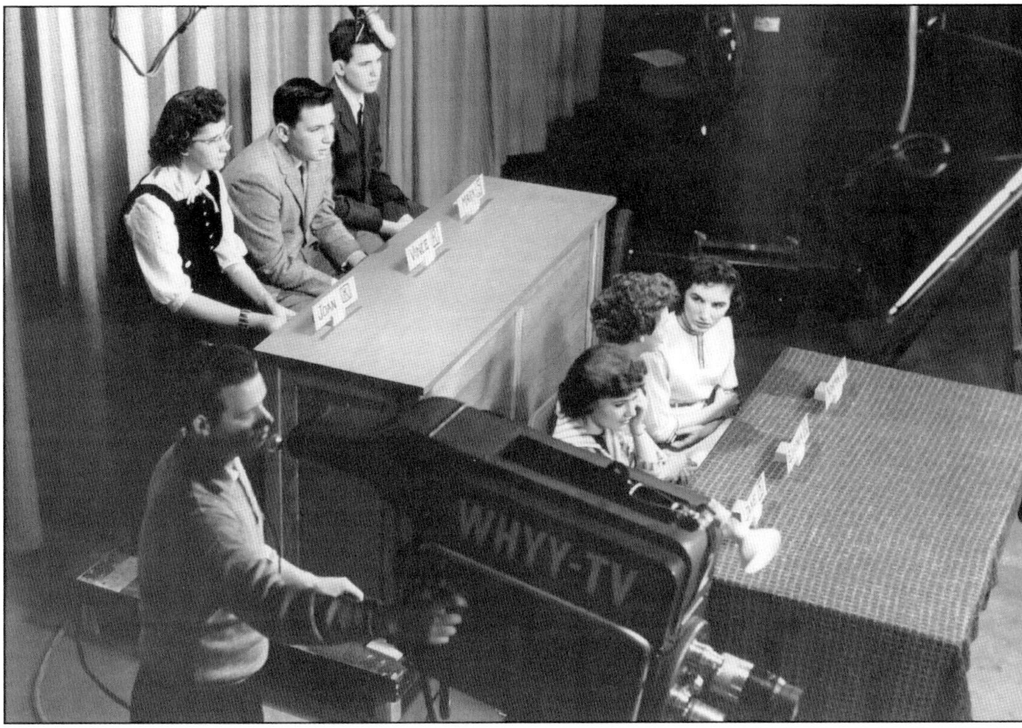

The Ed Goldman Show was a student quiz show that aired on WHYY in the late 1950s. The station's first home was 1622 Chestnut Street, the former headquarters of WCAU TV/Radio. (WHYY.)

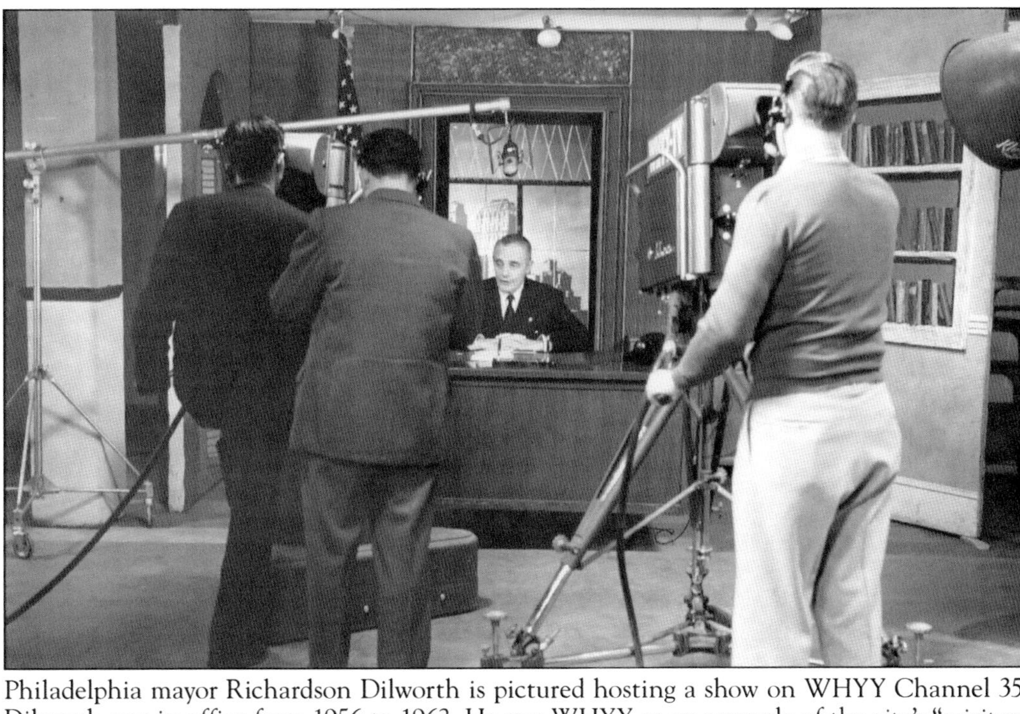

Philadelphia mayor Richardson Dilworth is pictured hosting a show on WHYY Channel 35. Dilworth was in office from 1956 to 1962. He saw WHYY as an example of the city's "spiritual awakening." (WHYY.)

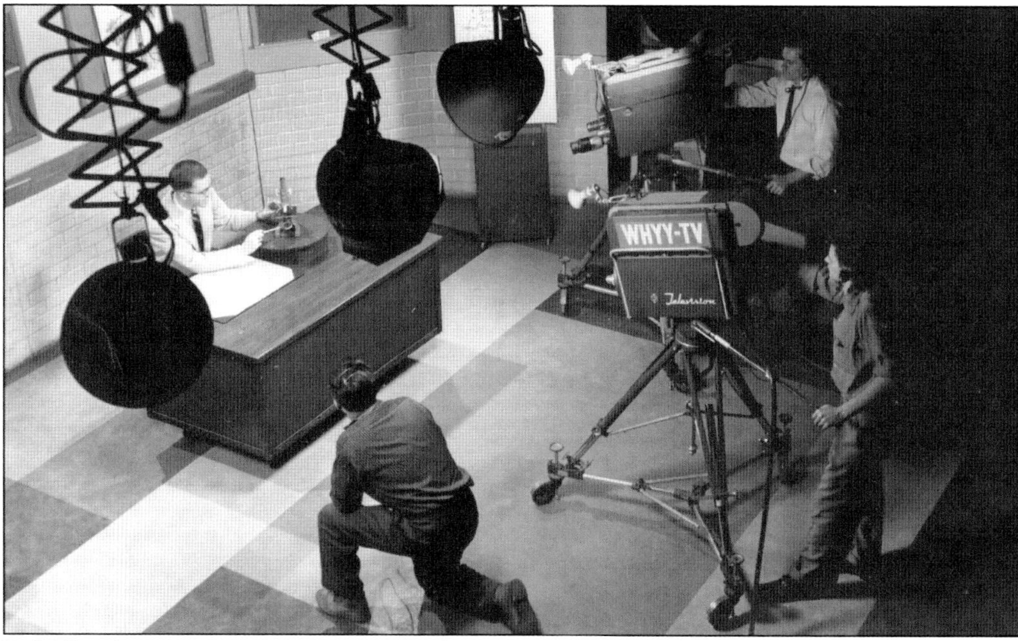

Bob Bloss hosts a program in WHYY's Chestnut Street studios on March 6, 1958. Bloss was the sports announcer for the station from 1957 to 1962. (WHYY.)

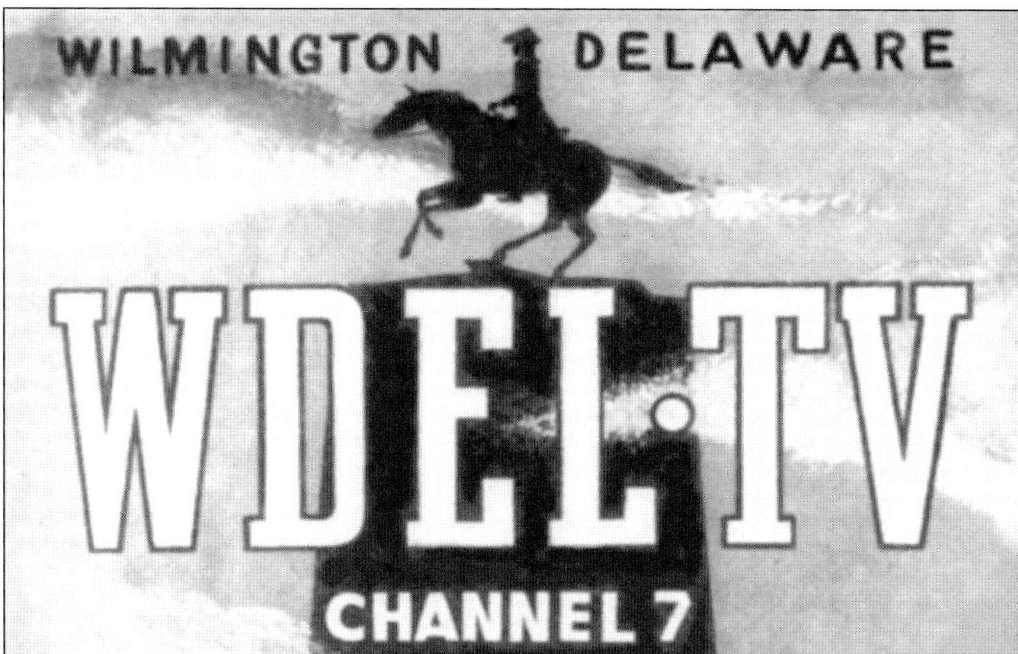

WHYY-TV can trace its history back to WDEL-TV. The Wilmington, Delaware, station was an NBC affiliate. It was on the air as Channel 7 from 1949 to 1951 and then moved to Channel 12 until 1958. The station went through several ownership changes, finally ending up with Storer Broadcasting. In a complex switch of licenses and deals with the FCC, Storer gave up its license for Channel 12. WHYY was granted permission to take over the unused dial position. The move gave WHYY a much better VHF signal that could be seen throughout the Philadelphia region. (WHYY.)

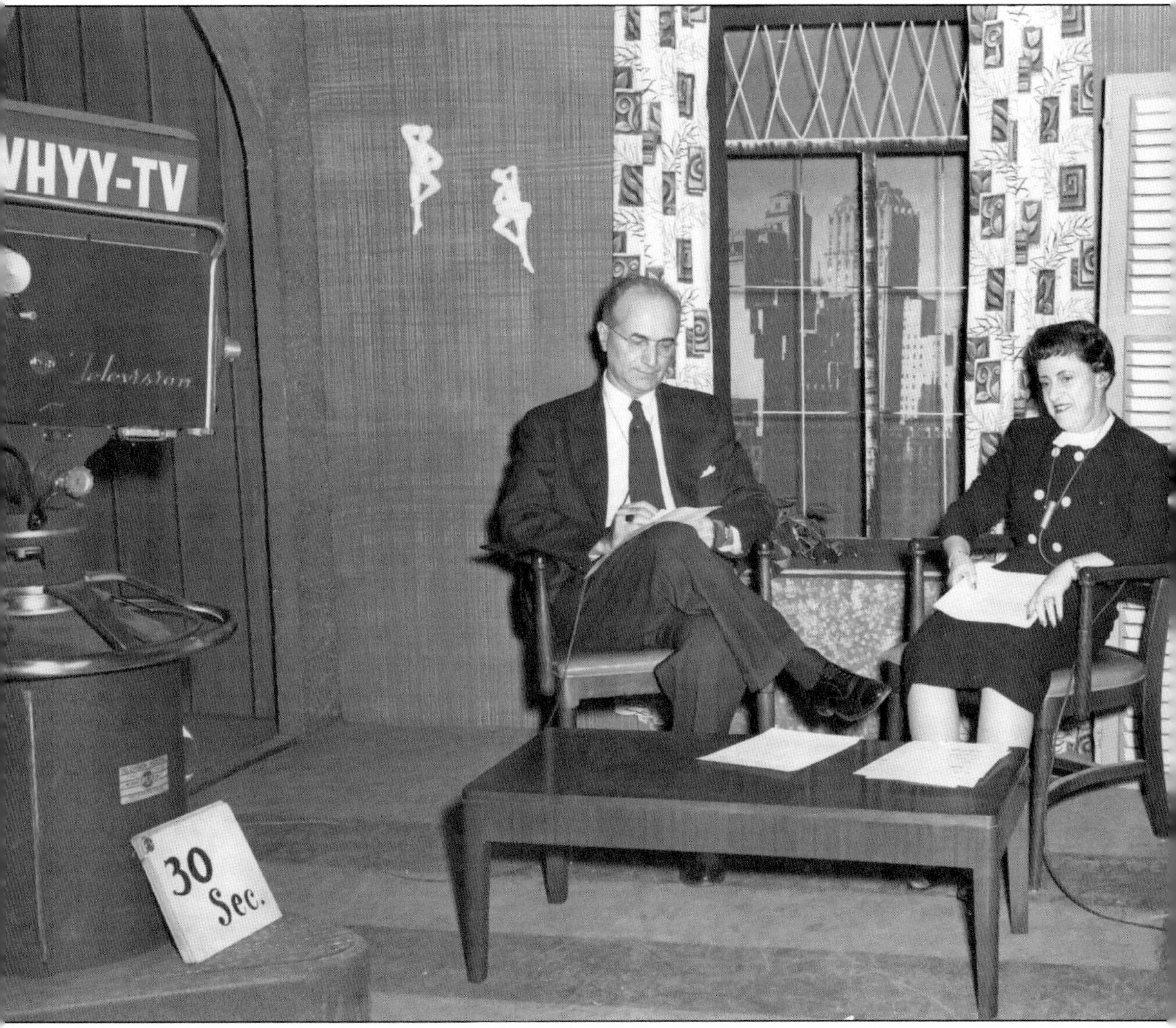

Pictured is a talk show in the WHYY studio in the 1950s. Notice the sign on the camera pedestal used to let the hosts know how much time is left in the show. (WHYY.)

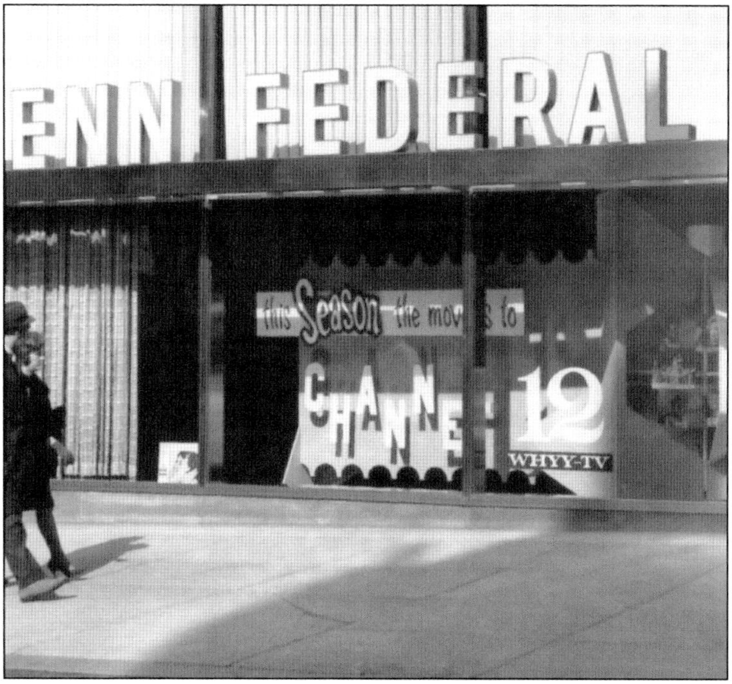

A sign in the window of a Center City storefront announces WHYY's move from Channel 35 to Channel 12. This was made possible when the FCC awarded WHYY the dial position of an abandoned license in Wilmington, Delaware. WHYY debuted on Channel 12 from a Wilmington studio on September 16, 1963. (WHYY.)

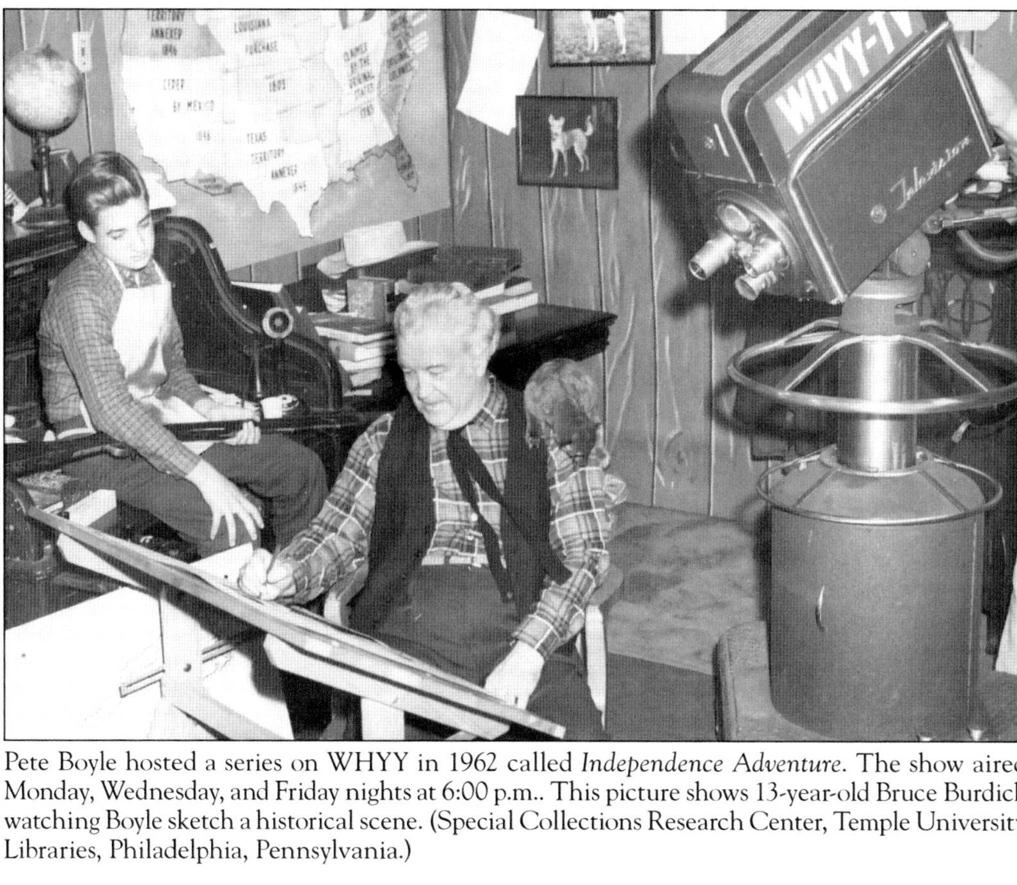

Pete Boyle hosted a series on WHYY in 1962 called *Independence Adventure*. The show aired Monday, Wednesday, and Friday nights at 6:00 p.m.. This picture shows 13-year-old Bruce Burdick watching Boyle sketch a historical scene. (Special Collections Research Center, Temple University Libraries, Philadelphia, Pennsylvania.)

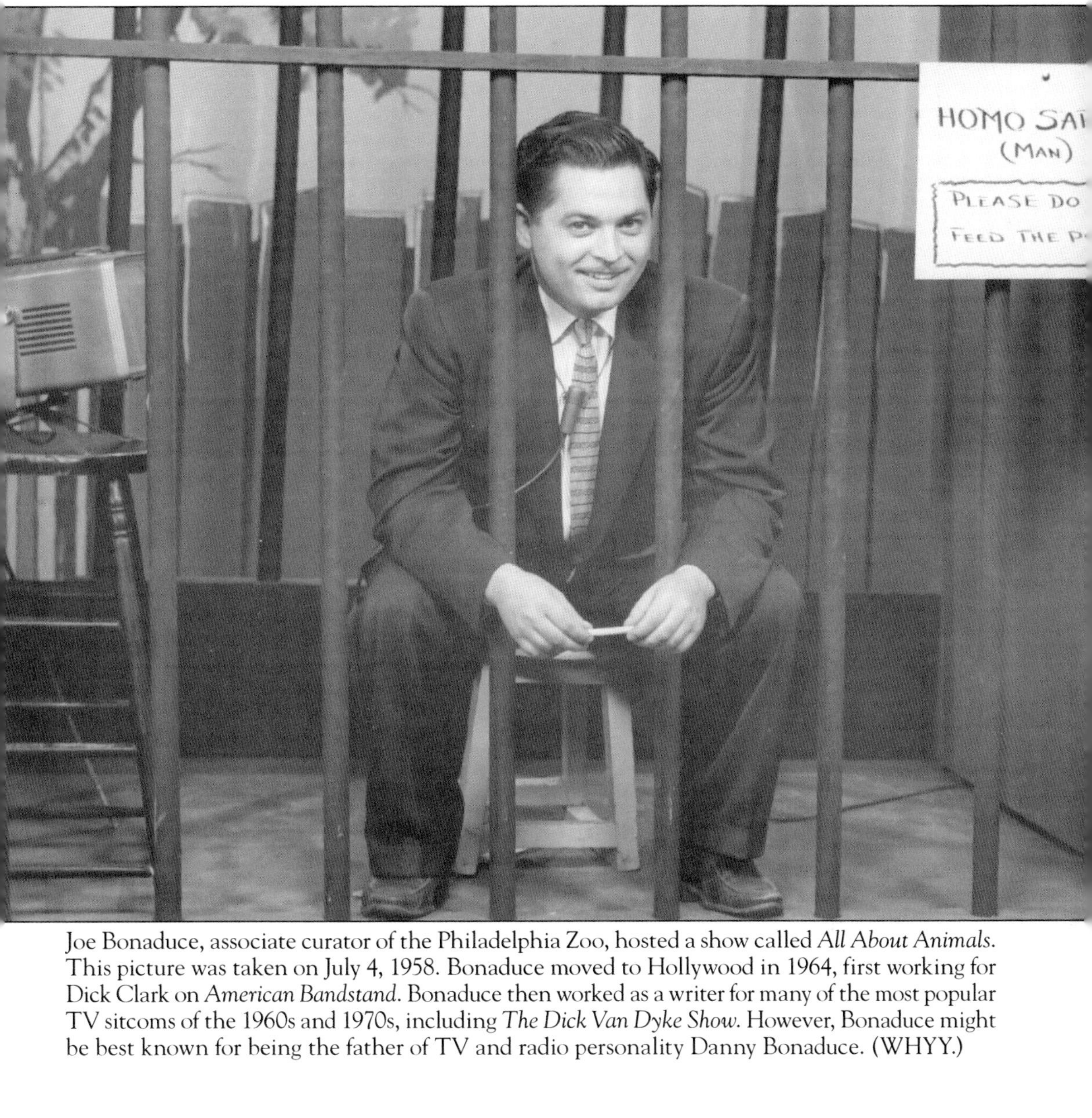

Joe Bonaduce, associate curator of the Philadelphia Zoo, hosted a show called *All About Animals*. This picture was taken on July 4, 1958. Bonaduce moved to Hollywood in 1964, first working for Dick Clark on *American Bandstand*. Bonaduce then worked as a writer for many of the most popular TV sitcoms of the 1960s and 1970s, including *The Dick Van Dyke Show*. However, Bonaduce might be best known for being the father of TV and radio personality Danny Bonaduce. (WHYY.)

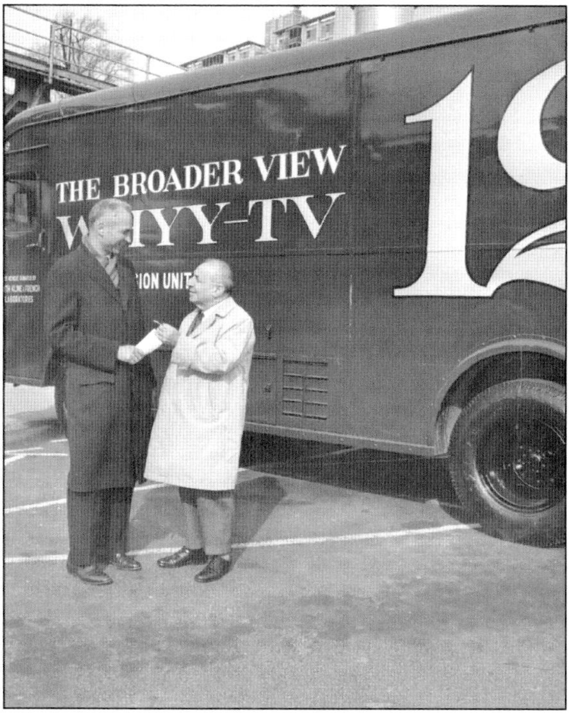

Frederick Osborn Jr., a former executive director of the Smith, Kline & French Foundation, hands over the keys to a new mobile broadcasting truck to WHYY president William Goldman in March 1964. The truck allowed the station to broadcast from remote locations. (Special Collections Research Center, Temple University Libraries, Philadelphia, Pennsylvania.)

Philadelphia mayor James Tate hands a $250,000 donation check to WHYY's William Goldman in the 1960s. Goldman was the Founder of WHYY and served as its president until his death in 1975. Goldman also owned a chain of movie theaters in the Philadelphia area. (WHYY.)

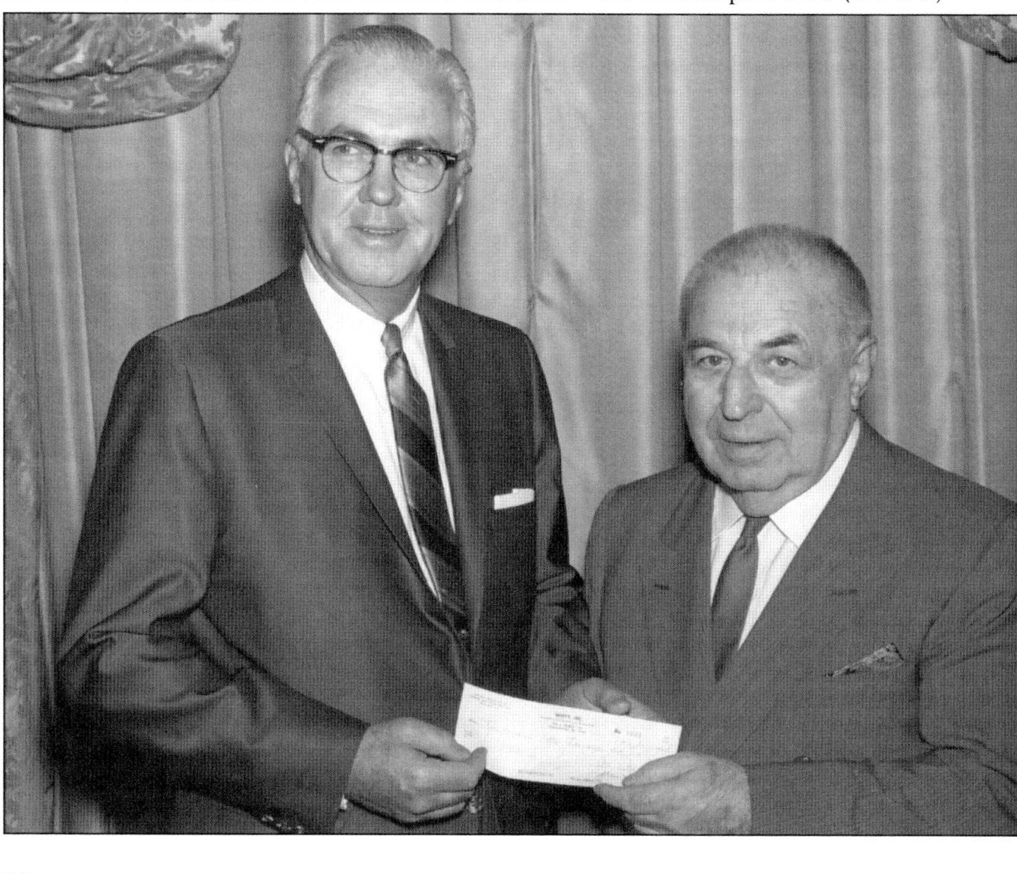

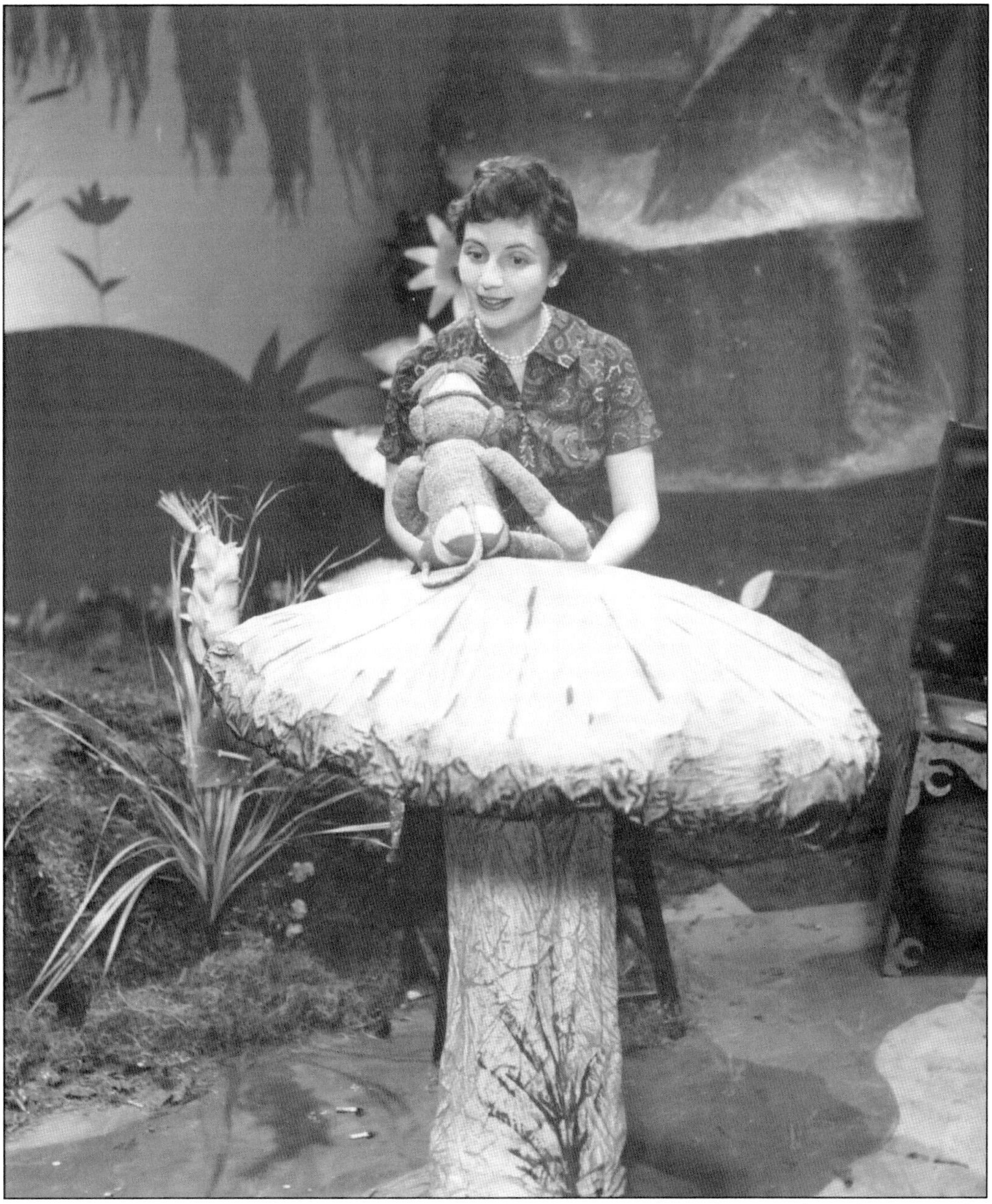

Host Kathy Molz is pictured in *The Land of Nod*, a children's show that aired for three months in the Summer of 1958.

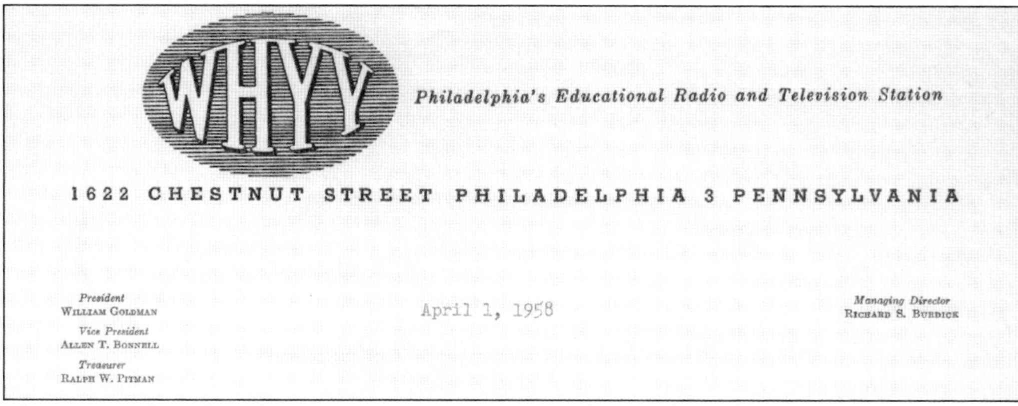

WHYY

Philadelphia's Educational Radio and Television Station

1622 CHESTNUT STREET PHILADELPHIA 3 PENNSYLVANIA

President
WILLIAM GOLDMAN

Vice President
ALLEN T. BONNELL

Treasurer
RALPH W. PITMAN

April 1, 1958

Managing Director
RICHARD S. BURDICK

This WHYY letterhead from 1958, when the station was still located at 1622 Chestnut Street, shows William Goldman as station president. (WHYY.)

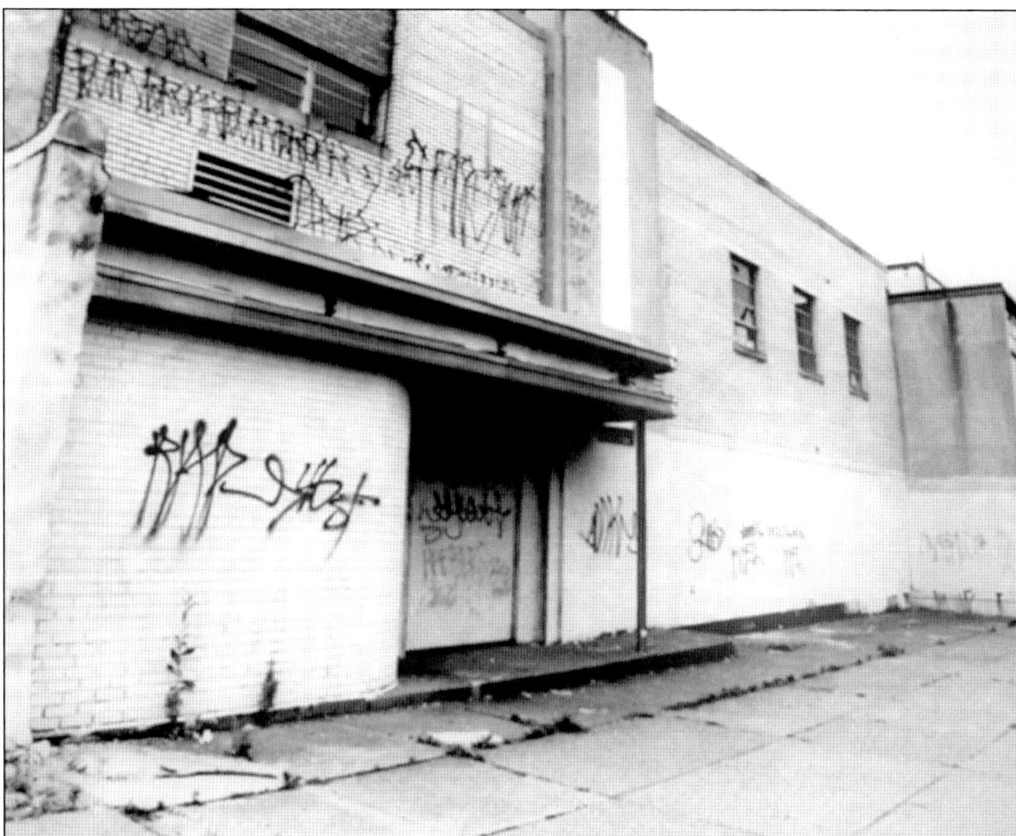

This is how the WHYY studios at 46th & Market Streets looked in the 1990s. The historic building was built by WFIL-TV and was the first home of *Bandstand*, but it fell into disrepair after WHYY moved out in 1979. It is now home to the Enterprise Center. (The Enterprise Center.)

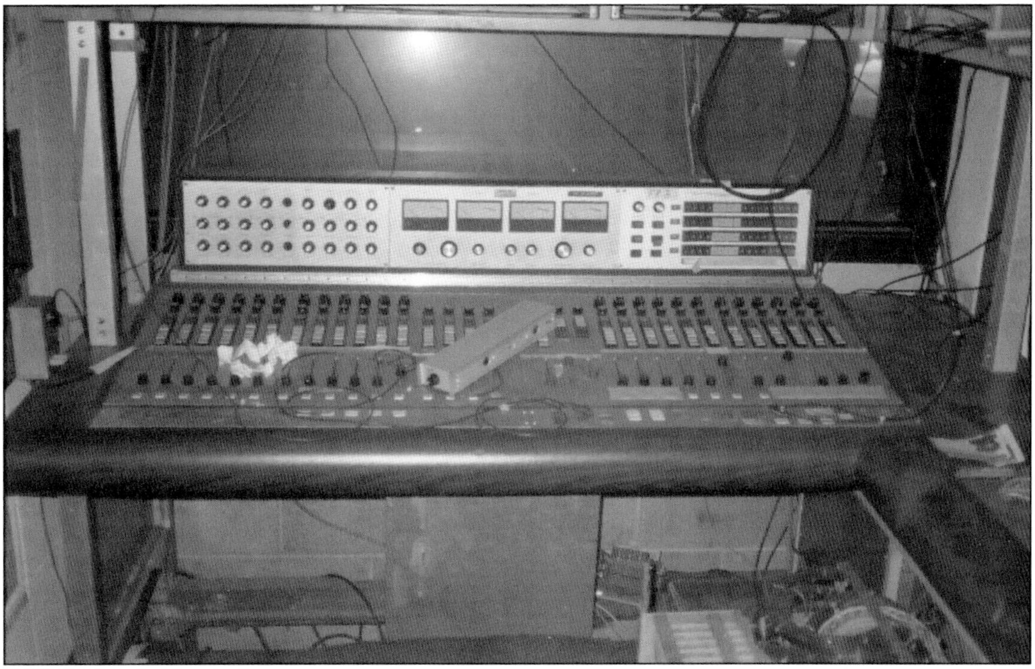

These two photographs show equipment inside the WHYY studios at Forty-sixth and Market Streets as the station packed up to move to its new building at Sixth and Race Streets. (Both, WHYY.)

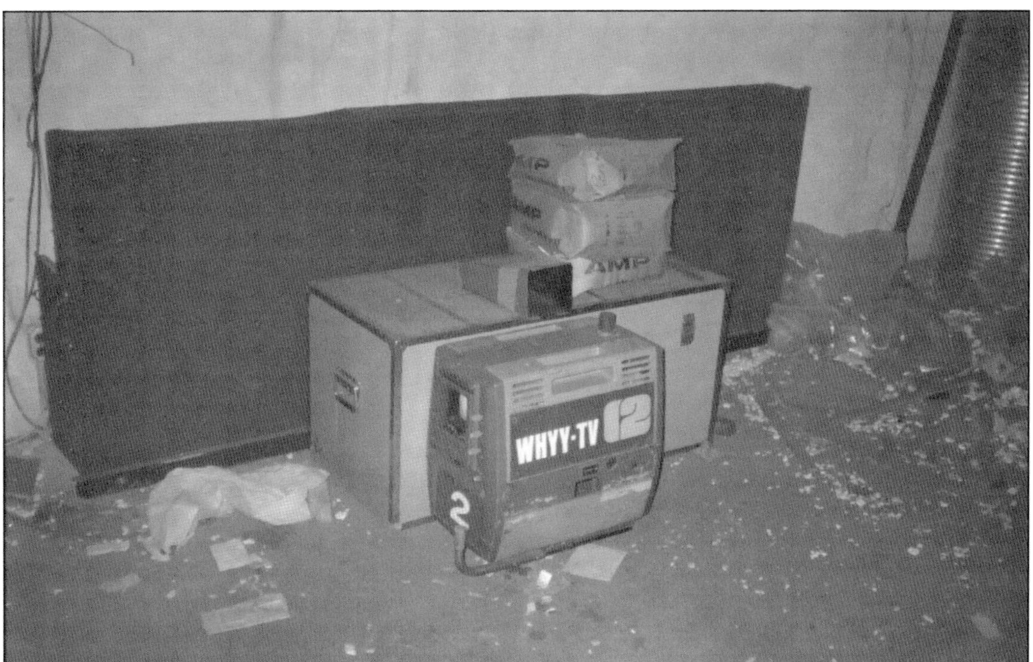

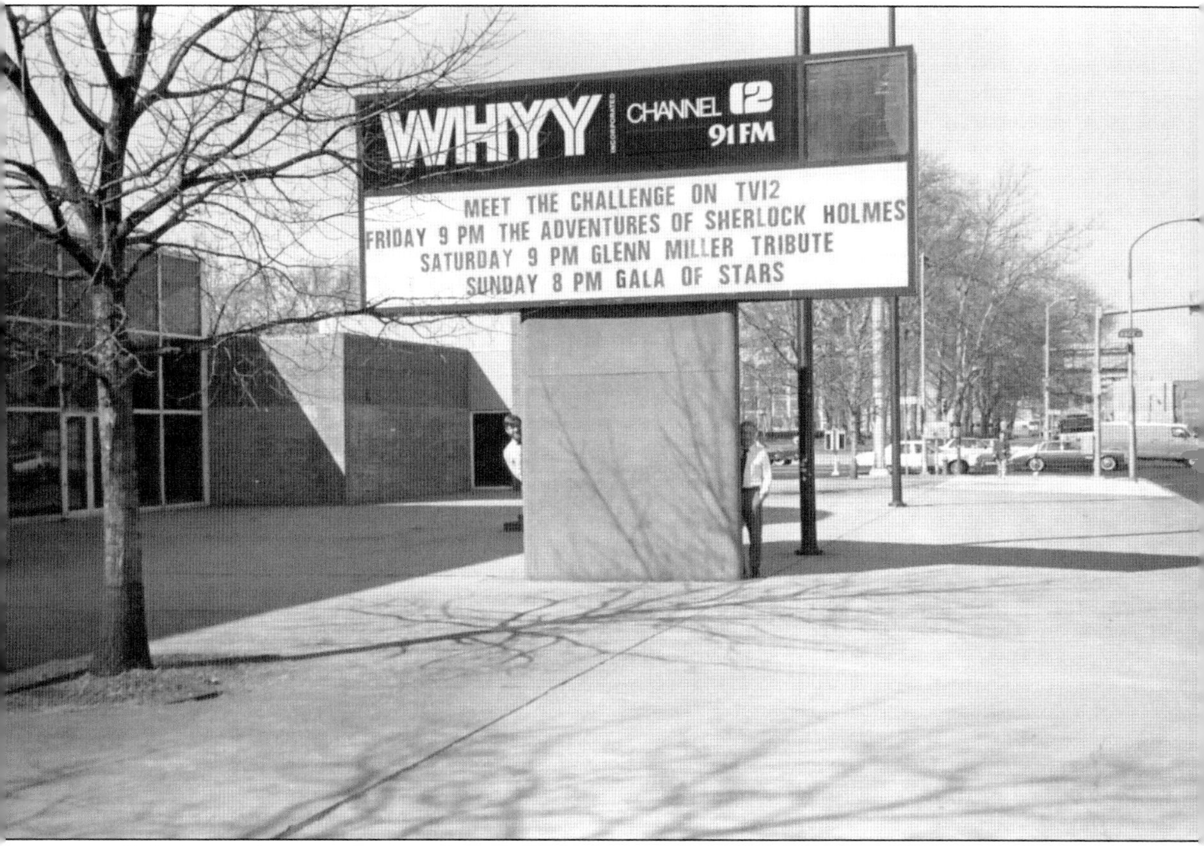

WHYY moved into its new (and current) home in February 1979. The building was the former Living History Center Museum at Sixth and Race Streets on the west side of Independence Mall. The museum opened for the city's Bicentennial celebration in 1976. WHYY-FM radio moved to the building about 18 months later. (WHYY.)

WHYY director of public information Barbara Leff looks for her chair after the PBS station moved from Forty-sixth and Market Streets to Sixth and Race Streets. (Special Collections Research Center, Temple University Libraries, Philadelphia, Pennsylvania.)

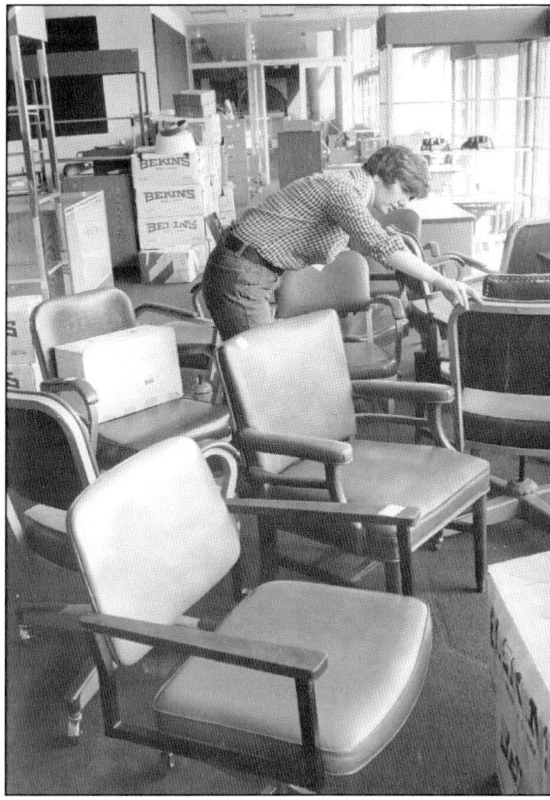

WHYY's Robert Hall (left) and an unidentified employee try to figure out what to unpack first at the former Living History Center Museum. Philadelphia mayor Frank Rizzo tried to block the move because he was upset at a documentary the station ran about him. Rizzo eventually gave in. (WHYY.)

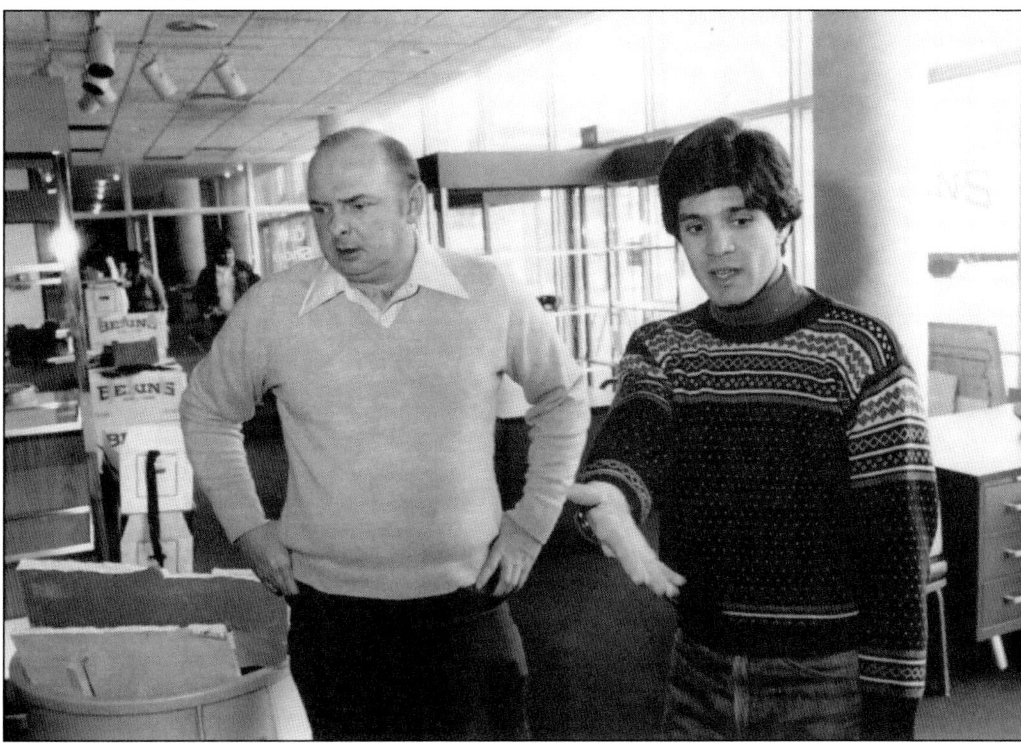

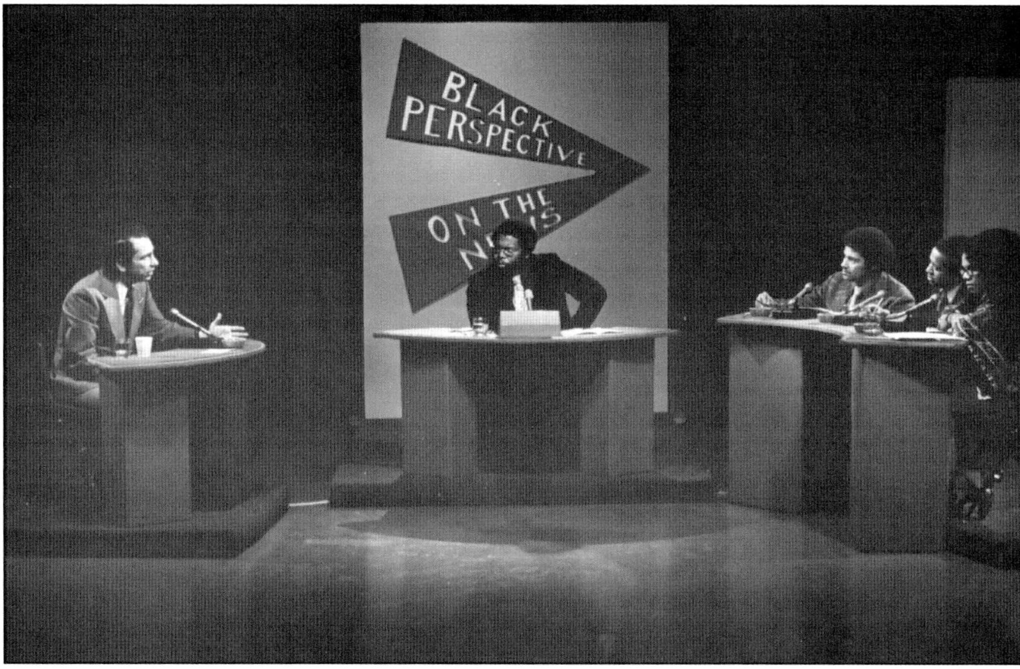

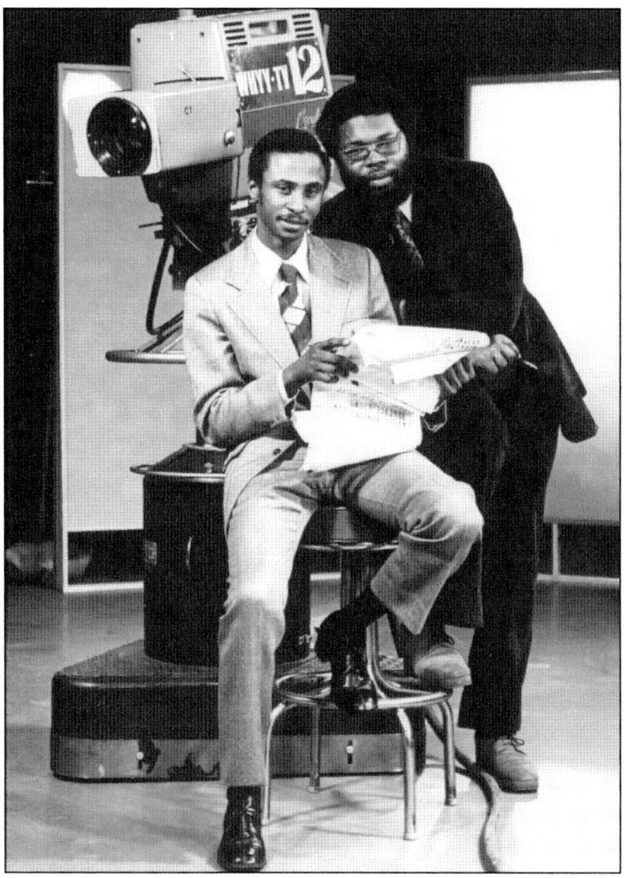

Black Perspective on the News aired on WHYY from 1973 to 1979. In this photograph from 1973, host Reggie Bryant (center) moderates a discussion with Philadelphia district attorney Arlen Specter and a panel of journalists. (WHYY.)

Cohosts Reggie Bryant (standing) and Acel Moore pose on the set of *Black Perspective on the News*. It was the first weekly series from WHYY to air nationally on the Public Broadcasting Service. (WHYY.)

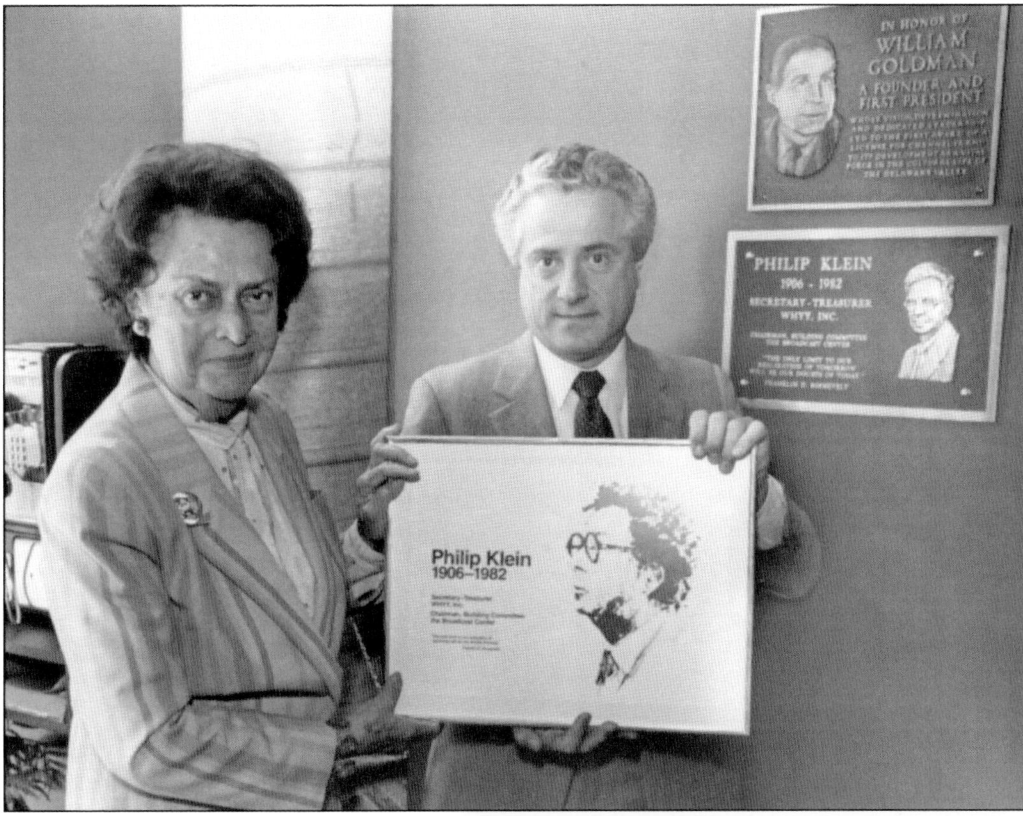

In 1983, Rick Breitenfeld helps unveil a plaque in honor of longtime station secretary-treasurer Philip Klein. It hangs in the WHYY lobby under the tribute to former president William Goldman. (WHYY.)

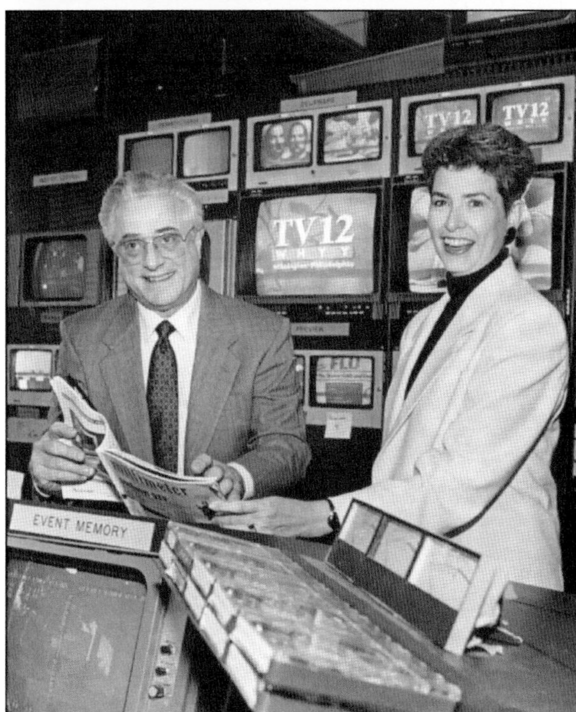

Breitenfeld and WHYY board member Lynn Herrick Sharp are pictured in one of the station's control rooms. (WHYY.)

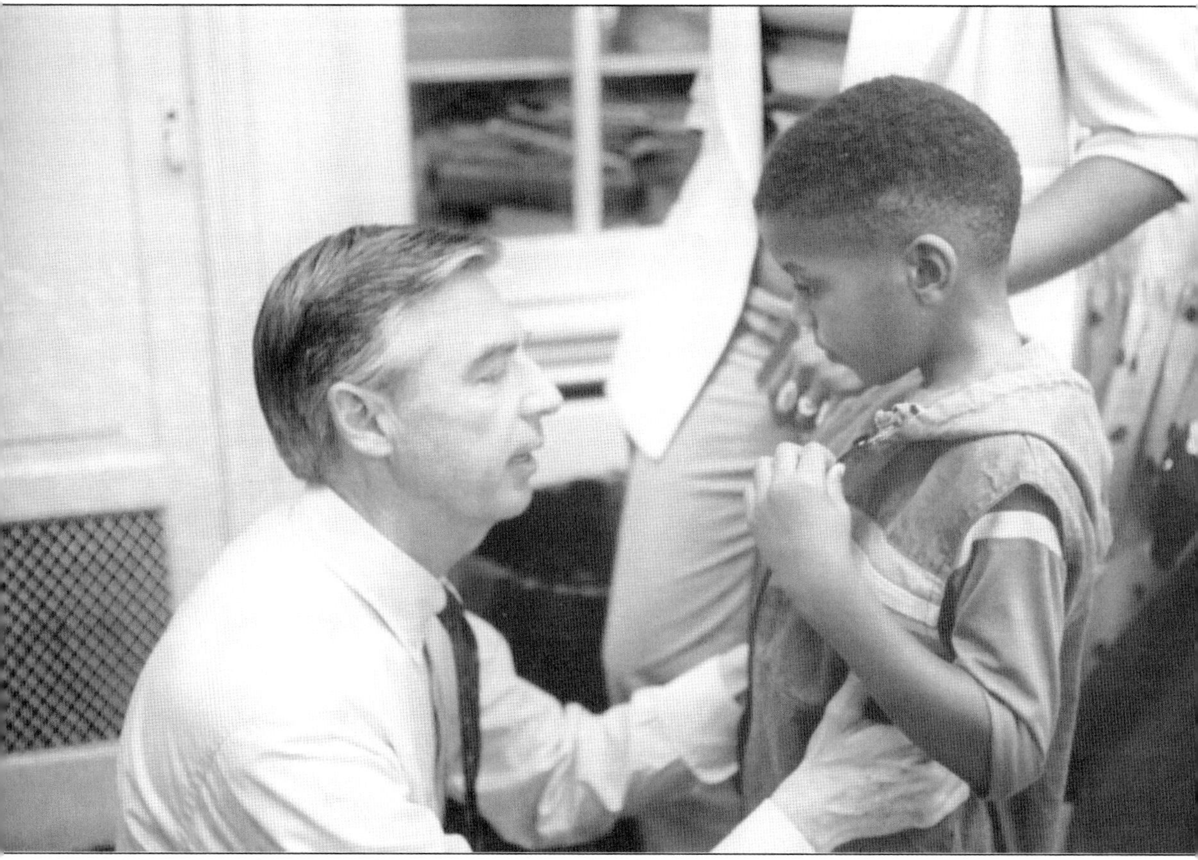

PBS icon Fred Rogers, of *Mister Rogers' Neighborhood*, paid a visit to Philadelphia in 1980. He visited local schools to help promote his children's show, which aired on WHYY. (WHYY.)

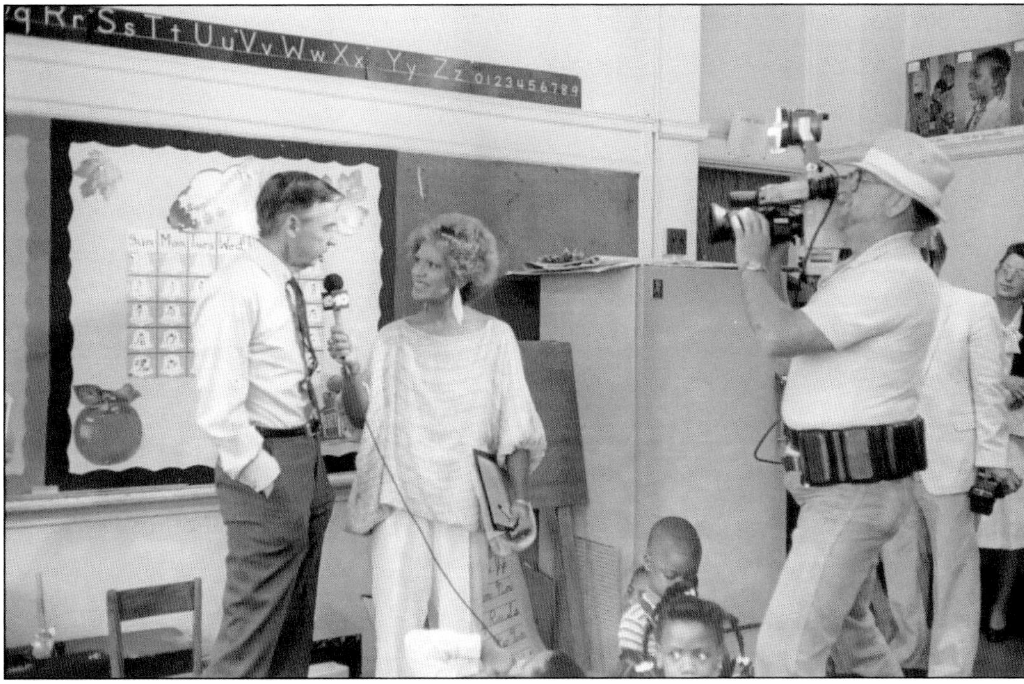

When Mr. Rogers comes to town, it is news. WCAU's Sheela Allen interviews the host at a Philadelphia school. Rogers produced his show at WQED in Pittsburgh from 1968 to 2001. (WHYY.)

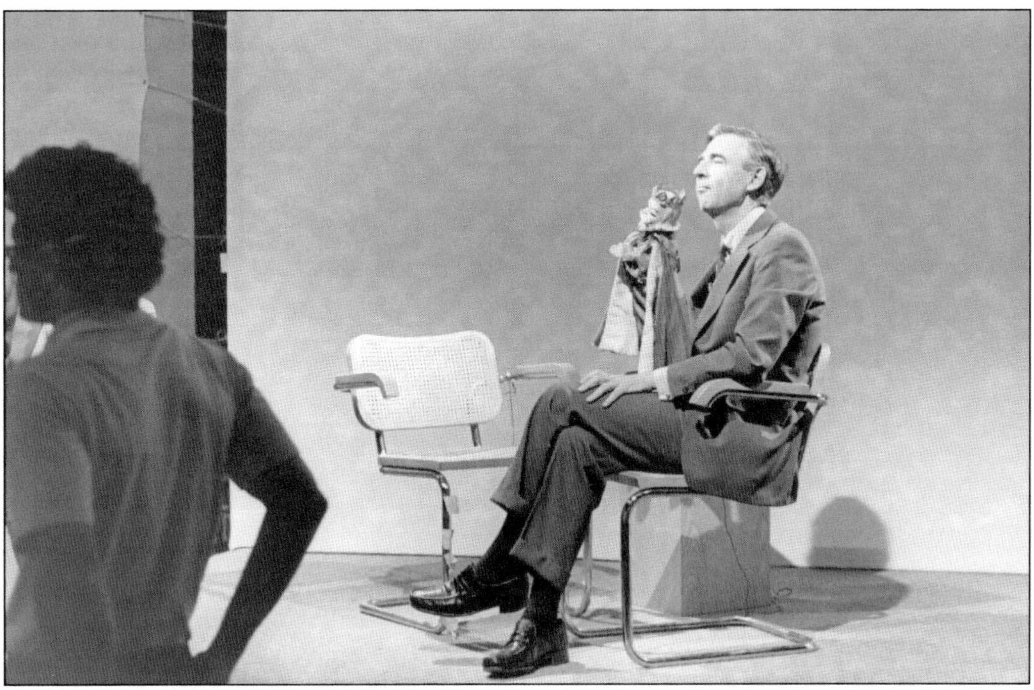

Mr. Rogers brought his "Neighborhood of Make Believe" to WHYY in 1980. Here, he demonstrates one of his puppet characters, King Friday XIII, in the studio. (WHYY.)

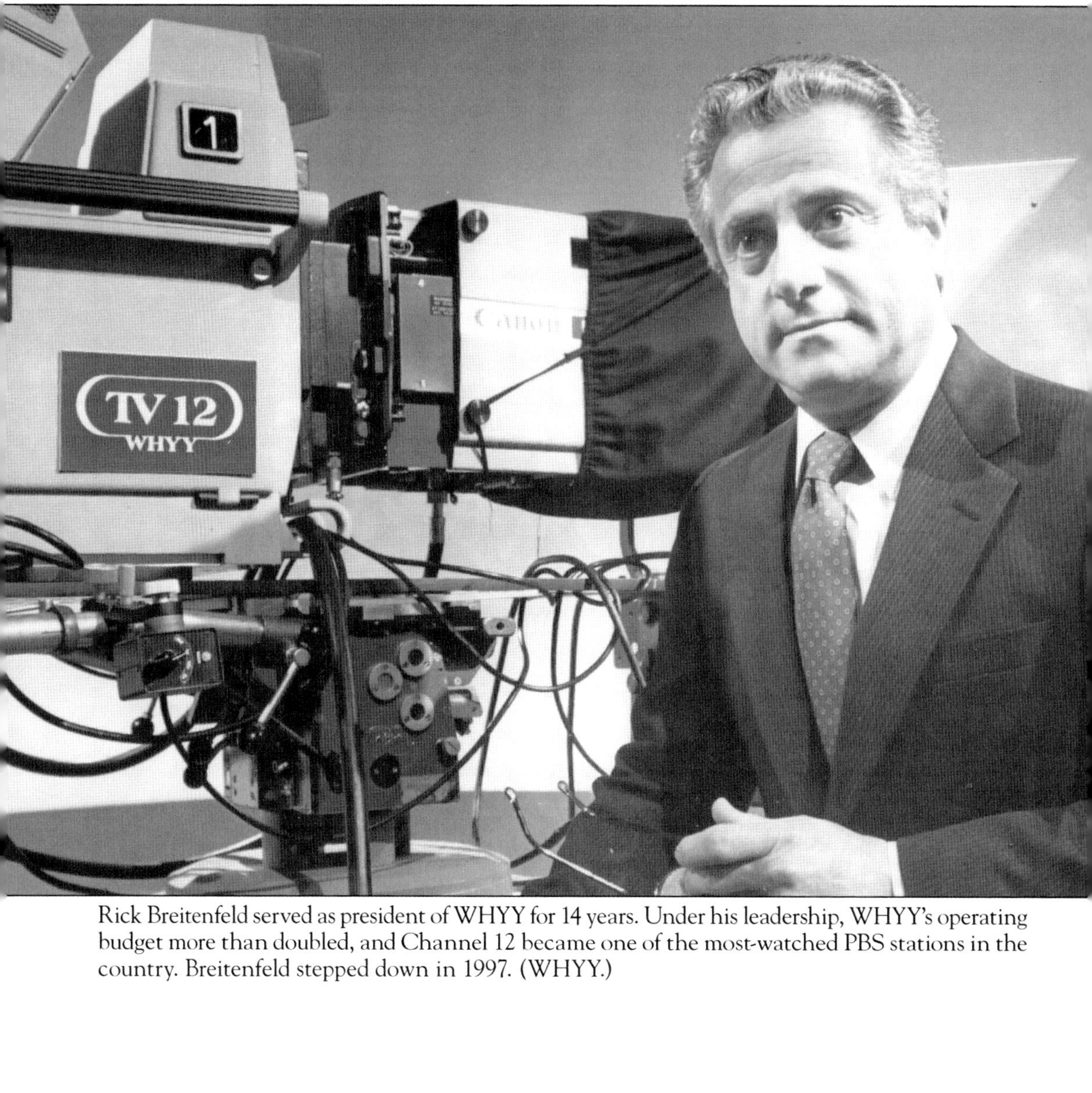

Rick Breitenfeld served as president of WHYY for 14 years. Under his leadership, WHYY's operating budget more than doubled, and Channel 12 became one of the most-watched PBS stations in the country. Breitenfeld stepped down in 1997. (WHYY.)

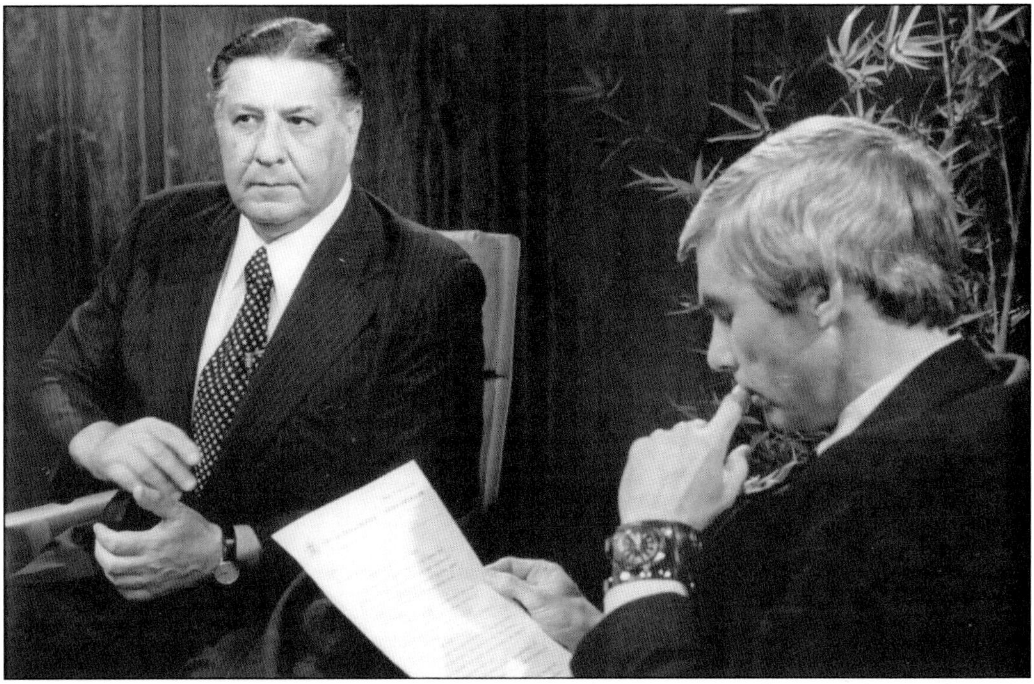

Philadelphia mayor Frank Rizzo had his differences with WHYY over the years, but sat down for an interview with the station in the late 1970s. (WHYY.)

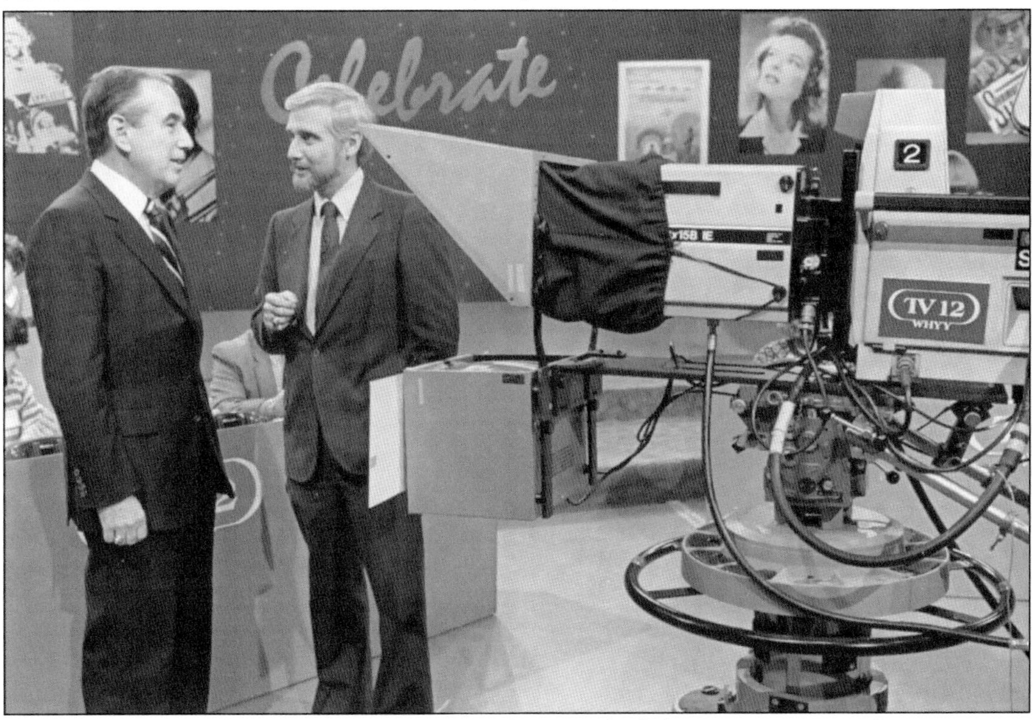

Fundraising is part of being a PBS station. Patrick Stoner (right) interviews a guest during a station pledge break in the mid-1980s. Stoner is an accomplished film critic and has been reviewing movies on WHYY for 30 years. (WHYY.)

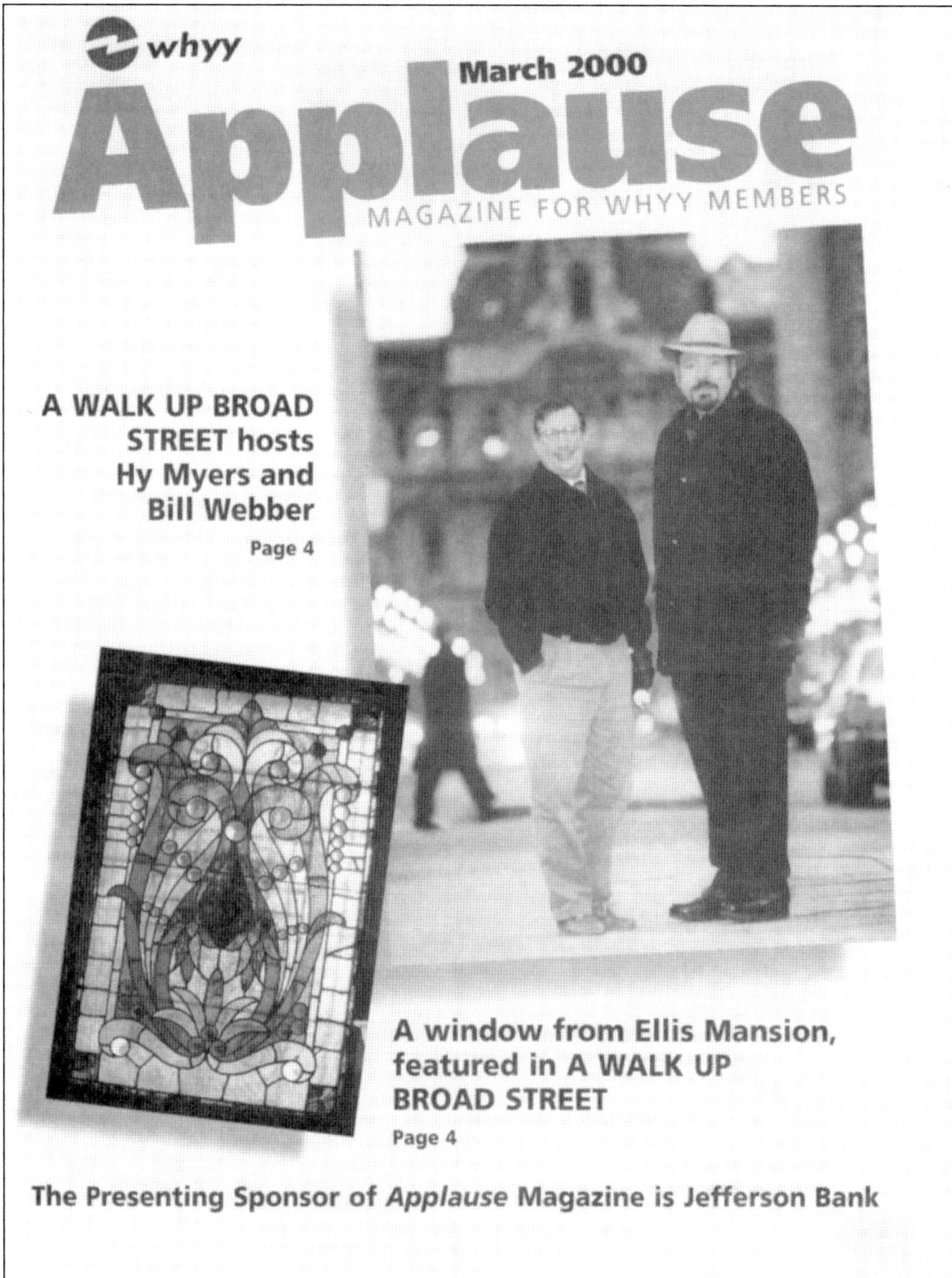

Broadcasting legend Bill Webber ended his long career at WHYY. In 2000, Webber (right) joined co-host Hy Myers for *A Walk Up Broad Street*. The documentary profiled Philadelphia's famous avenue that runs for 13 miles from the Navy Yard in South Philly to Cheltenham. The show was produced by WHYY's Ed Cunningham. (WHYY.)

Five

UHF Channels 17, 29, and 48

UHF TV had been around since the 1950s, but got a big boost in 1965 when the FCC ordered that all TV sets must include channels 13 through 72. Three Philadelphia stations jumped into the mix right away. WIBF-TV Channel 29 was first on the air in the summer of 1965, followed by WKBS-TV Channel 48 and WPHL-TV Channel 17 in September. The lack of major network commitments allowed all three stations to pick up local sports teams. Channel 17 was the home of the Philadelphia Phillies from 1965 to 2014 (with a brief stint on Channel 29 in the 1980s). Philadelphia Flyers games have aired on Channels 29 and 17. The 76ers have been on all three stations at various times. UHF stations have been very popular with younger viewers, focusing on children's shows and reruns of network sitcoms.

The three major UHF stations have all aired 10 o'clock newscasts. Channel 48 was the first, hitting the prime time airwaves in 1968. WTXF Channel 29 was next with *The 10 O'Clock News* in 1986. The first anchors were Lee McCarthy and Jill Chernekoff, with Howard Eskin on sports and Marie Michelini on weather. WPHL launched its own prime-time newscast in 1994 in conjunction with the *Philadelphia Inquirer. The Inquirer News Tonight* was anchored by Jim Watkins and Victoria Hong. The station dropped the *Inquirer* and produced its own newscast until 2005. WCAU produced the news for WPHL until 2012, when WPVI took over. Several alumni of Philadelphia UHF stations have continued their careers on national programs, including Jenna Wolfe (former sportscaster at WPHL) and Sheinelle Jones (former host of *Good Day Philadelphia* on WTXF), who are both hosts and reporters on NBC's *Today* (also known as *The Today Show*).

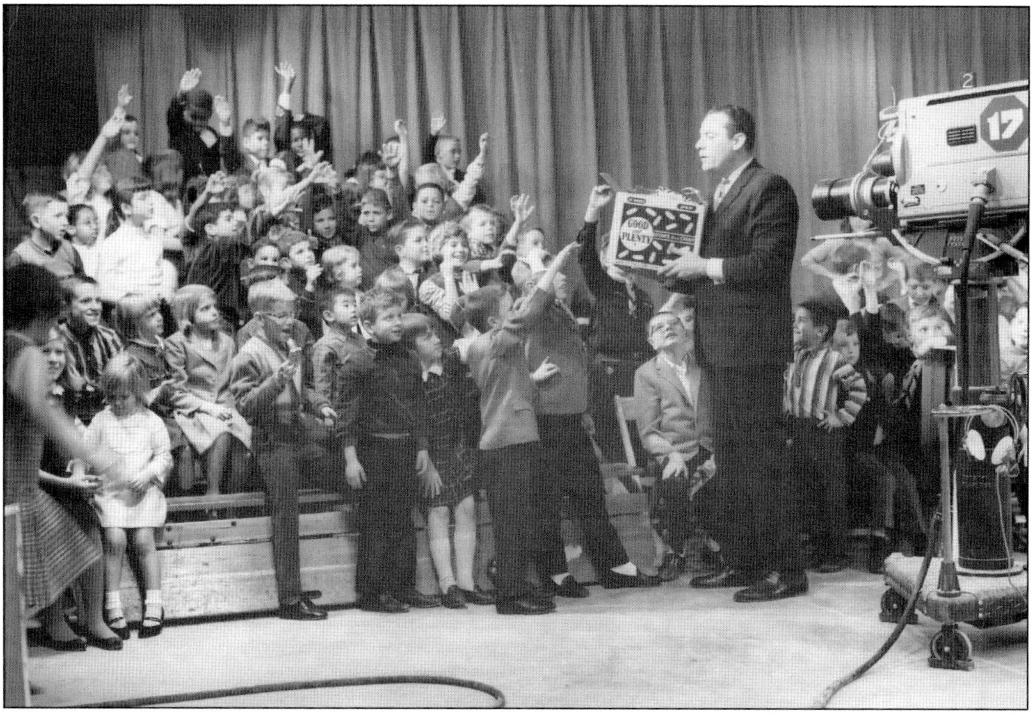

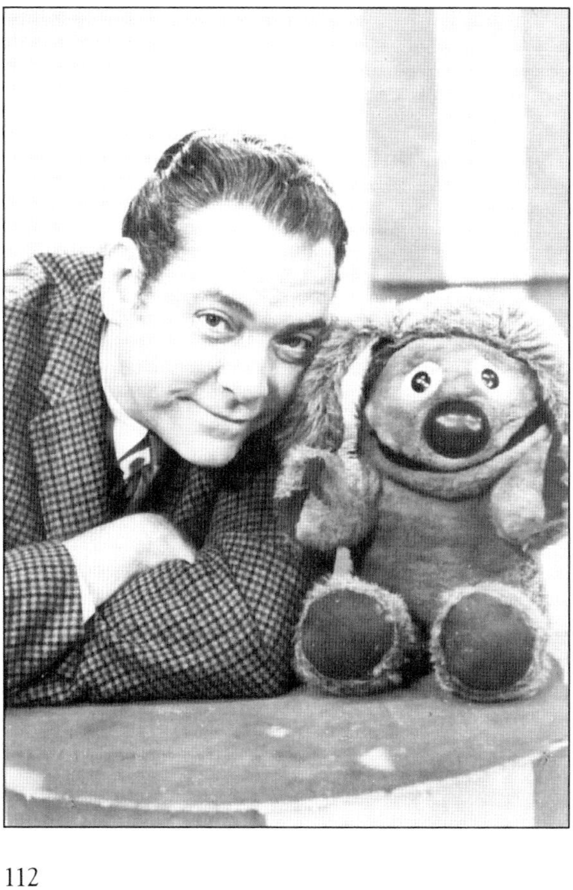

Bill "Wee Willie" Webber hosted the *Colorful Cartoon Club* from 1965 to 1975. This photograph shows Webber in the "Peanut Gallery" of local children who were a big part of the show. Webber also hosted an afternoon children's show on WKBS-TV from 1976 to 1979. (Special Collections Research Center, Temple University Libraries, Philadelphia, Pennsylvania.)

Webber featured puppet friends on his Channel 17 show, including Ralph, pictured here. Webber was also known for his midday air shift on WIP-AM radio. (Bill Webber Jr.)

WPHL's first studios were at its transmitter site on Mermaid Lane in Wyndmoor. In 1975, Channel 17 moved to this old A&P supermarket on Wynnefield Avenue in Philadelphia. The station was still there in 2015. (WPHL-TV.)

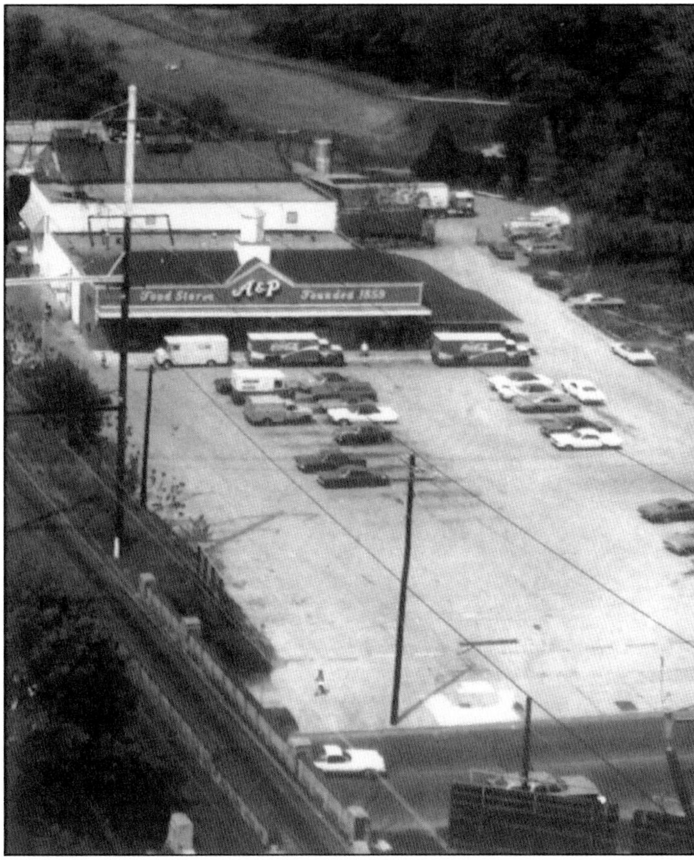

"Television on the Move" was the slogan on the side of WPHL's mobile broadcasting truck from the 1960s. (WPHL-TV.)

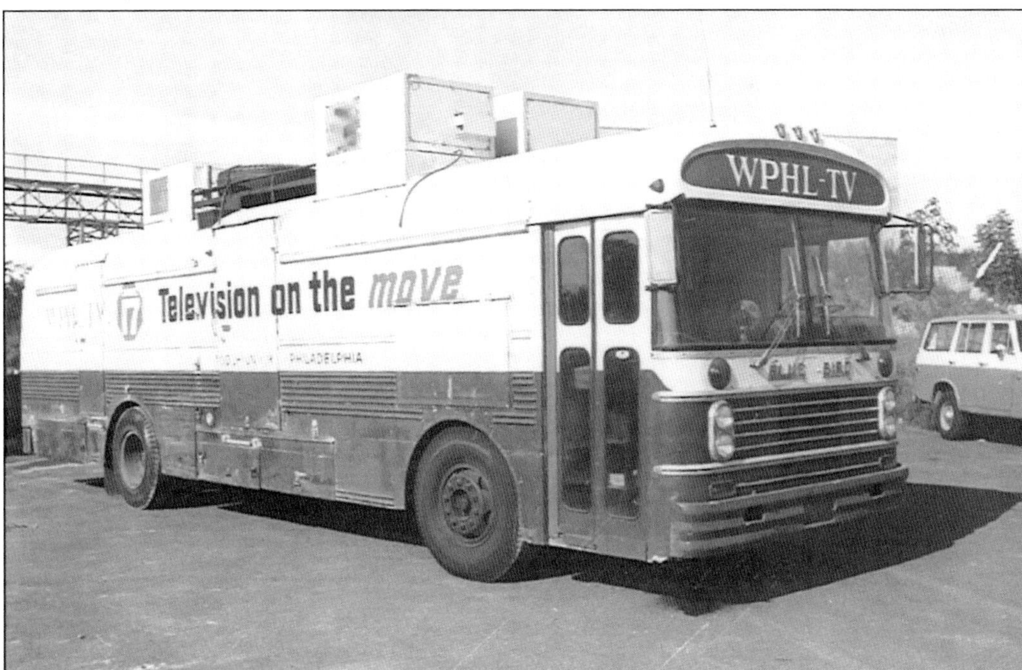

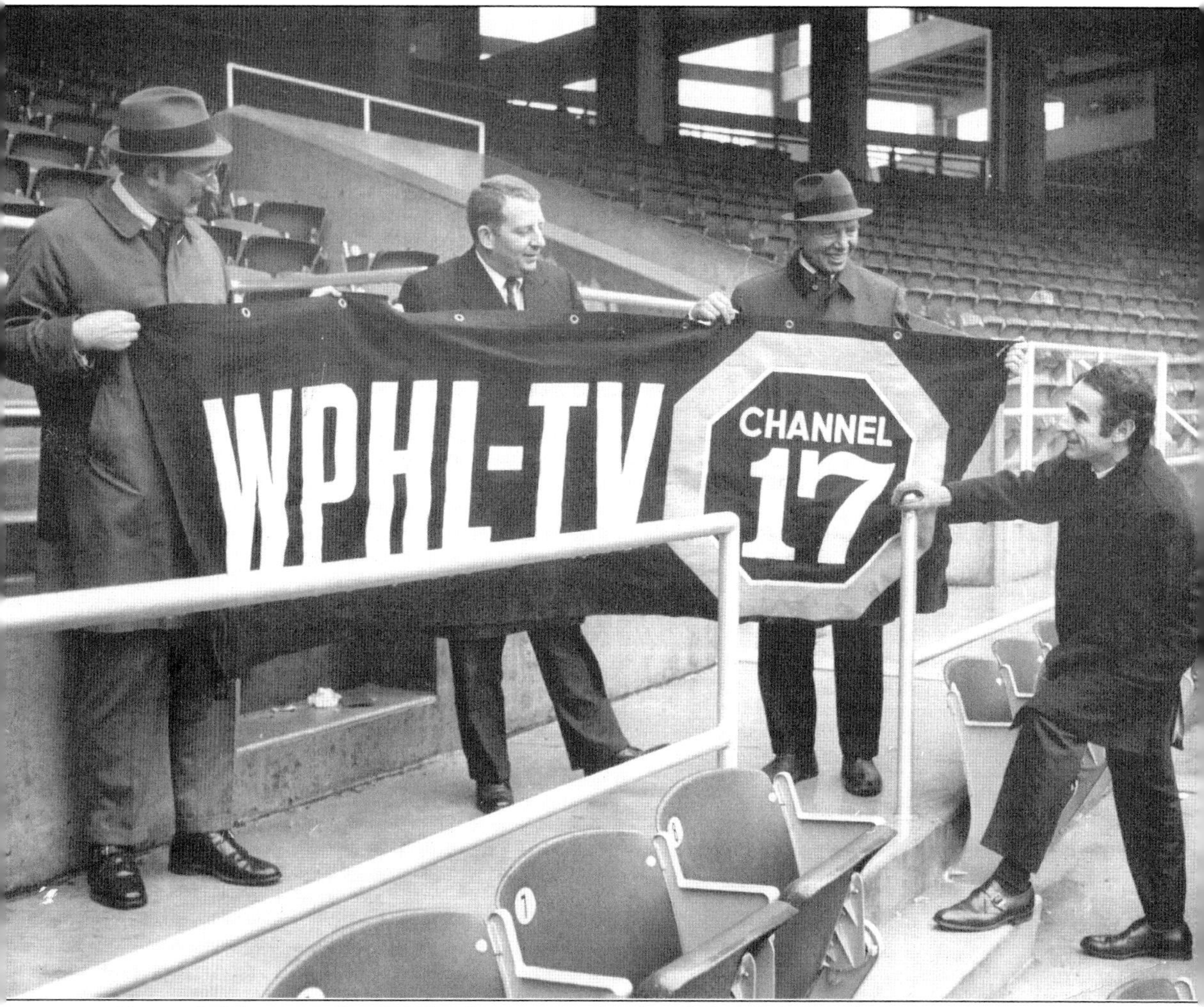

Channel 17 became the TV flagship station for the Philadelphia Phillies in 1971. This picture was taken in the "new" Veteran's Stadium. From left to right are Phillies vice president (and future owner) Sam Feinberg, Bill Giles, an unidentified executive, and Channel 17 founder Len Stevens. (WPHL-TV.)

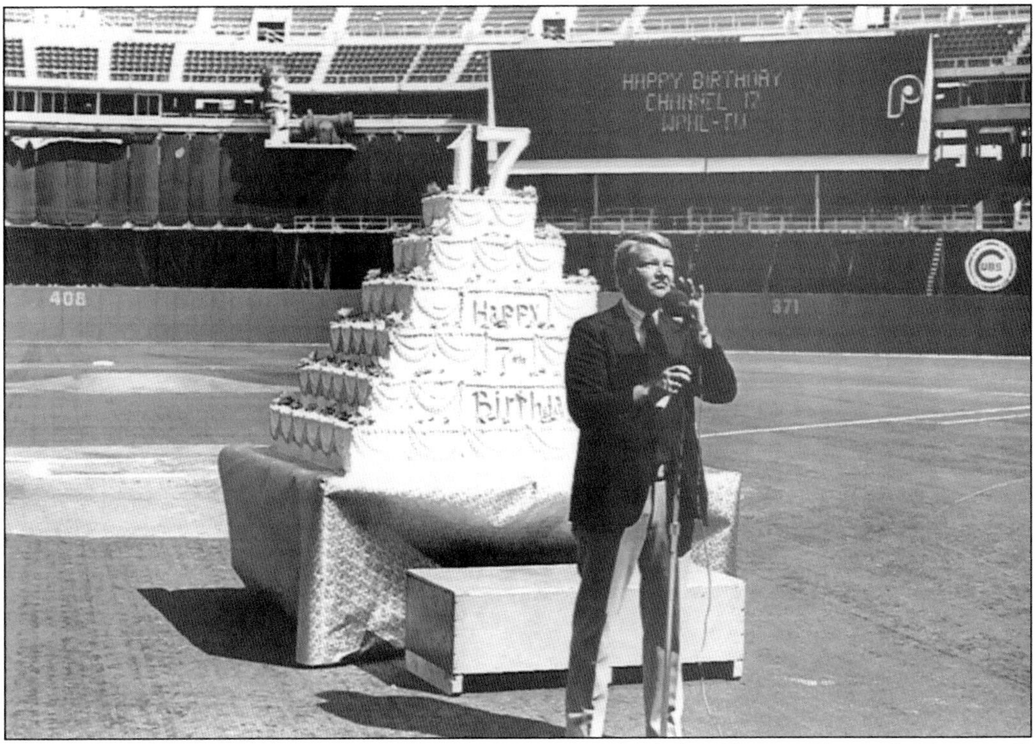

The Phillies' legendary play-by-play announcer Harry Kalas helped honor Channel 17's seventh birthday with a large cake on the field at Veteran's Stadium. Channel 17 was the home of the Phillies from 1971 to 1982, 1993 to 1998, and in conjunction with Comcast Sportsnet until 2014. (WPHL-TV.)

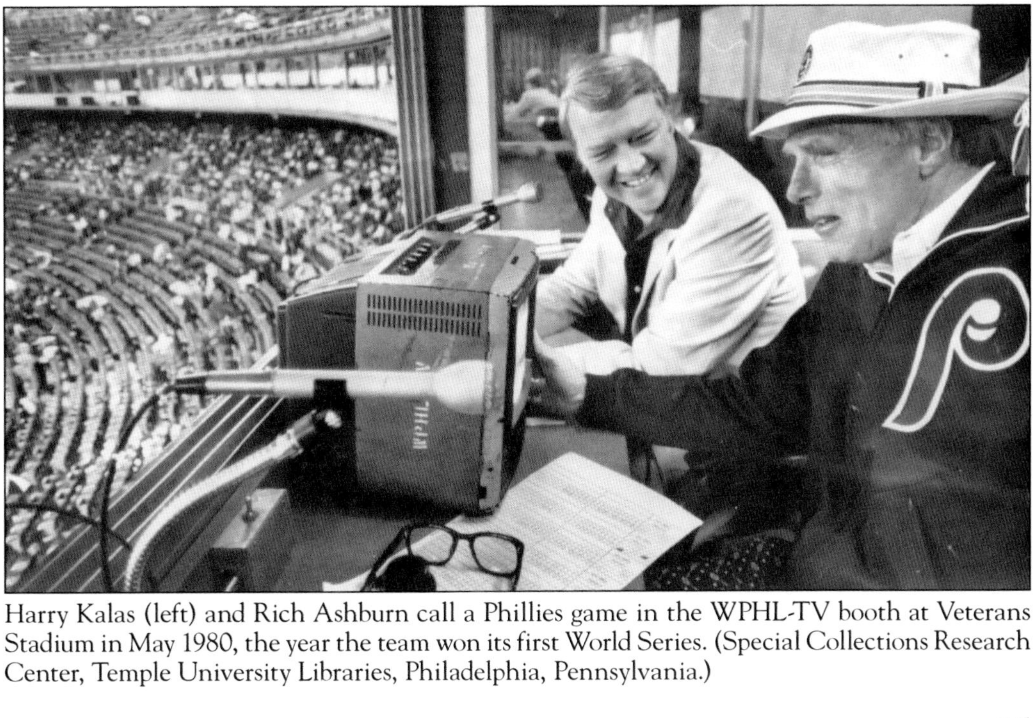

Harry Kalas (left) and Rich Ashburn call a Phillies game in the WPHL-TV booth at Veterans Stadium in May 1980, the year the team won its first World Series. (Special Collections Research Center, Temple University Libraries, Philadelphia, Pennsylvania.)

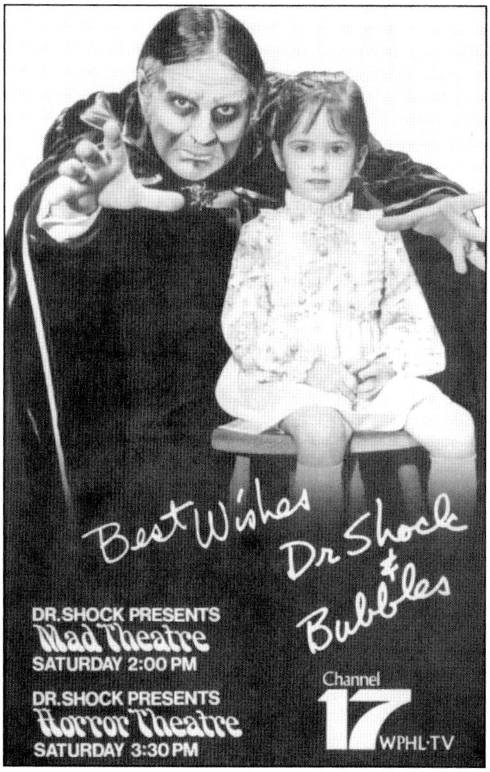

Dr. Shock was the most popular personality on WPHL. He hosted *Mad Theater* and *Horror Theater* from 1969 to 1979. The real name of "Doc Shock" was Joseph Zawislak. His real daughter Doreen was known as Bubbles on the shows. (Special Collections Research Center, Temple University Libraries, Philadelphia, Pennsylvania.)

Doctor Shock's sign-off, "Let there be fright," became a mantra for viewers. Zawislak died of a heart attack in 1979 at age 42. (Special Collections Research Center, Temple University Libraries, Philadelphia, Pennsylvania.)

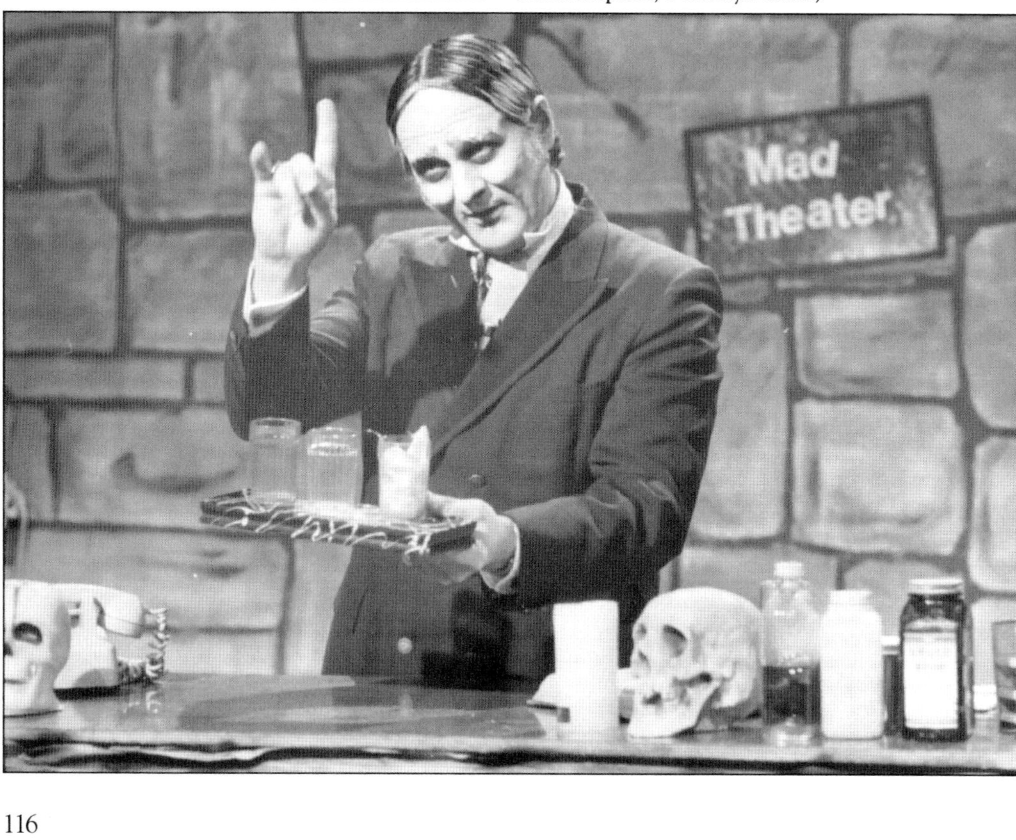

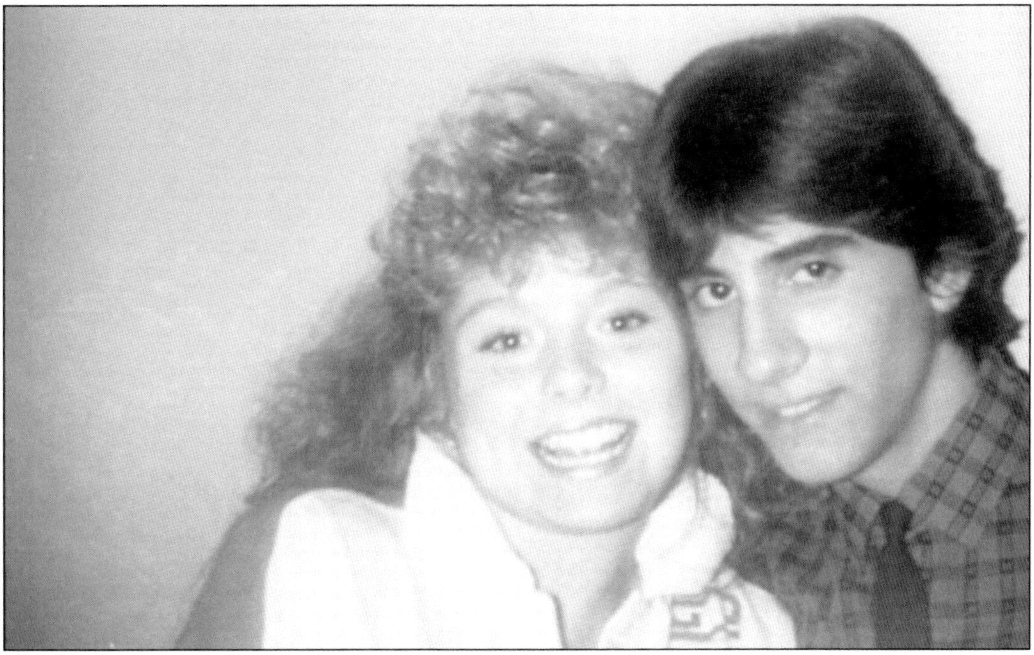

"Kelly from Voorhees," a popular regular on Channel 17's *Dancin' On Air*, was Kelly Ripa, who made her television debut on the after-school dance party (seen here with one of her dance partners). Ripa used the show as a stepping-stone to a role *on All My Children* and, eventually, her own morning talk show on ABC. (WPHL-TV.)

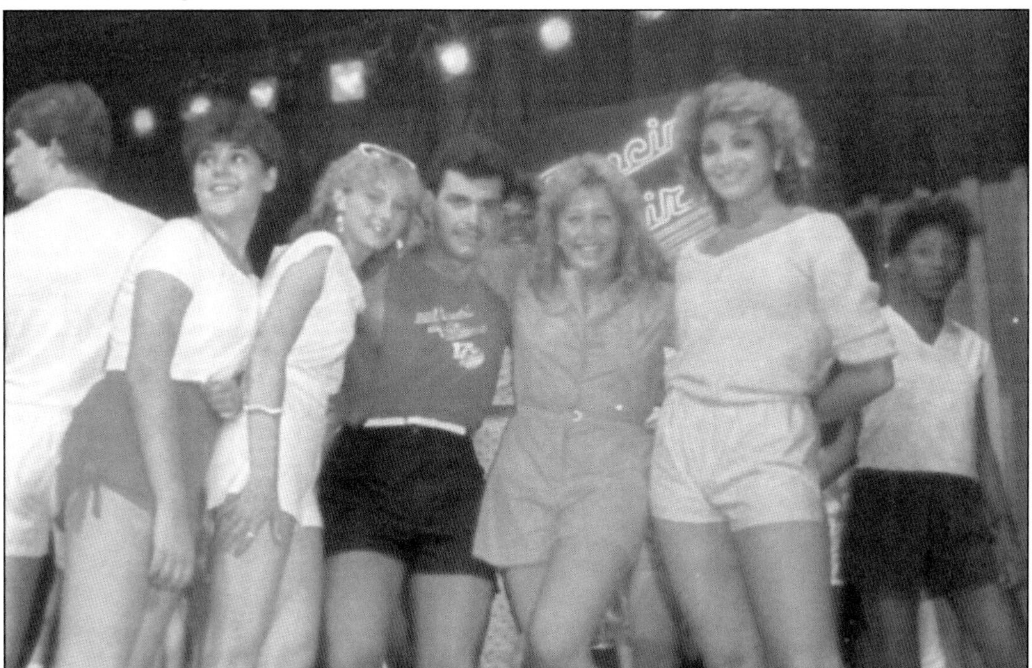

This is another photograph of a young Kelly Ripa (second from right) with some of the *Dancin' On Air* dancers. Eddie Bruce hosted the show from 1981 to 1987. Mike Nise was the producer. Madonna and Duran Duran made their American television debuts on the show. (WPHL-TV.)

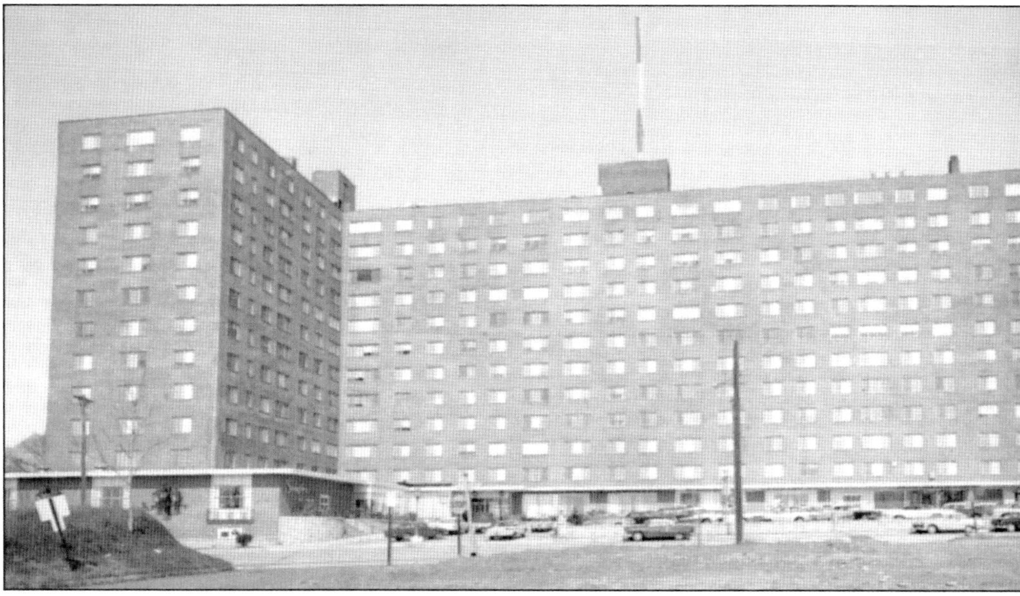

Channel 29's first home was the Benson East apartment building at Township Line and Old York Roads in Jenkintown. The transmitter was on the roof, with the studios in the basement. The station moved to its current location at Fourth and Market Streets in Old City in the early 1970s. (Author's collection.)

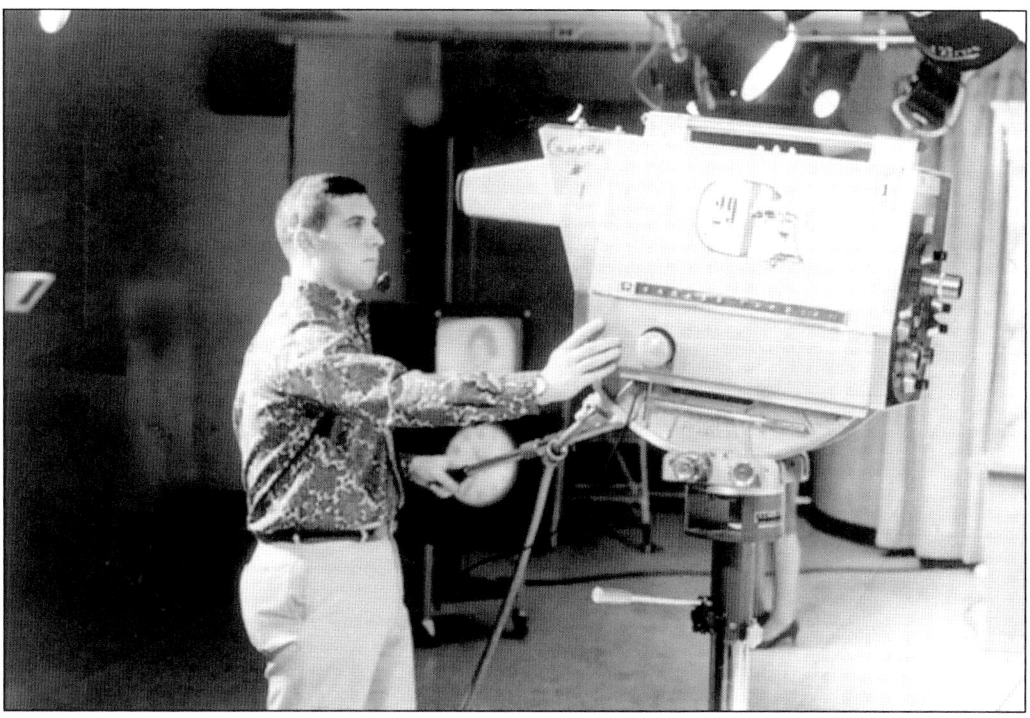

Bob Fleischer runs the camera in the WIBF Benson East studio in 1966. (Bob Fleischer.)

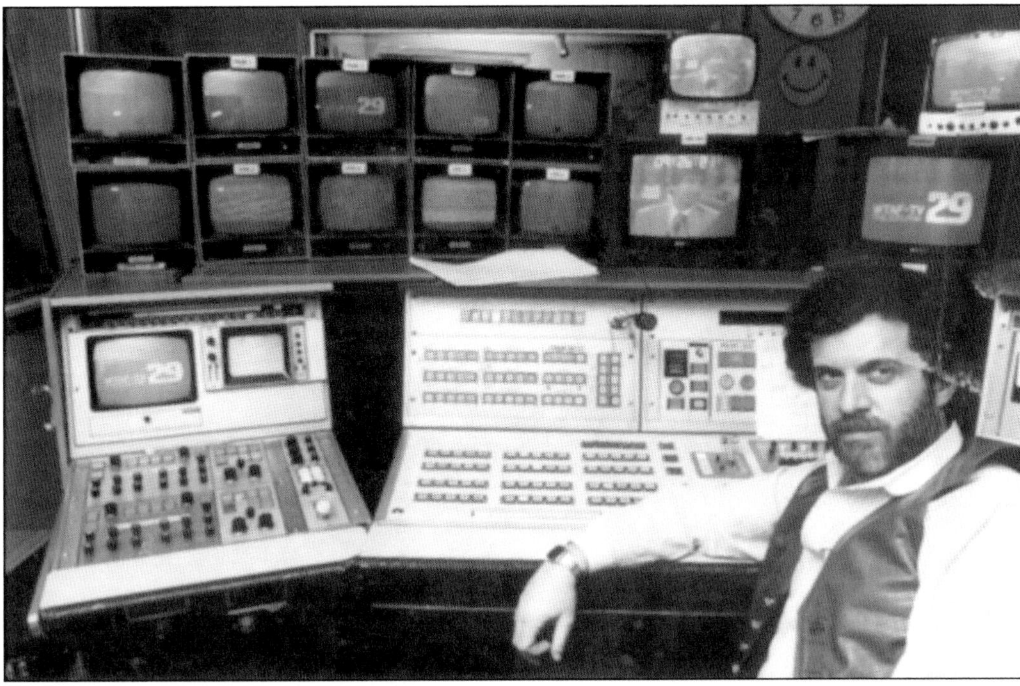

Bob Fleischer sits in the station's master control room in the Fourth and Market Streets studio (when the call letters were still WTAF). Fleischer was among those who helped news director Roger LeMay launch the *The 10 O'Clock News* on February 1, 1986. (Bob Fleischer.)

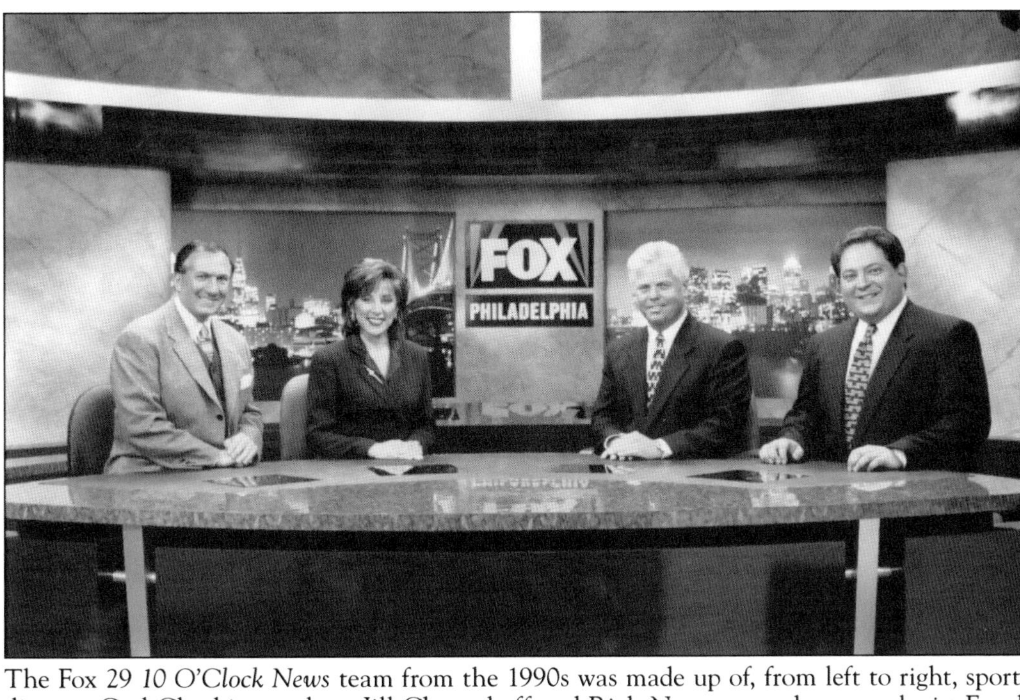

The Fox 29 *10 O'Clock News* team from the 1990s was made up of, from left to right, sports director Carl Cherkin, anchors Jill Chernekoff and Rich Noonan, and meteorologist Frank Cariello. (WTXF-TV.)

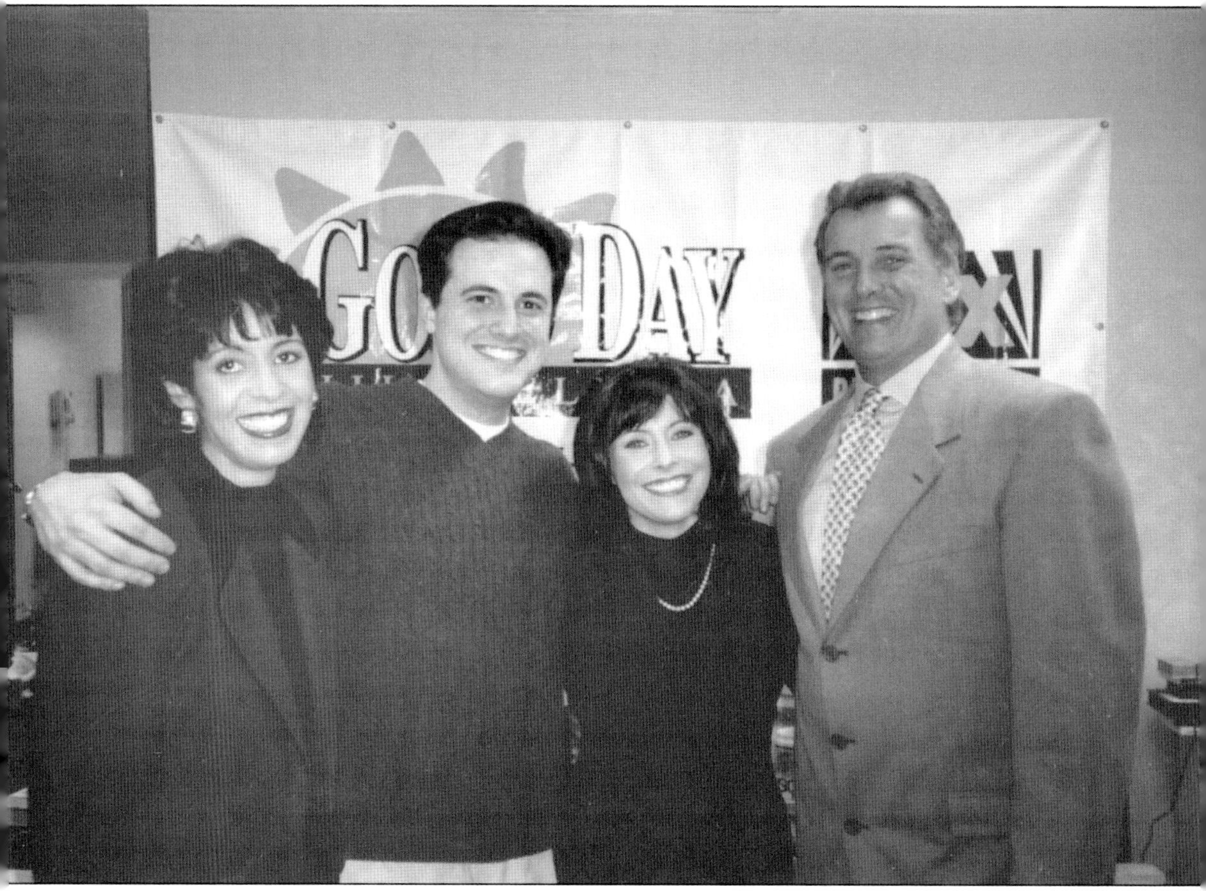

On April 1, 1996, WTXF-TV debuted *Good Day Philadelphia*, a three-hour, locally produced morning show. Pictured from left to right are news anchor Tracy Matisak, producer Bob Bicknell, and hosts Donya Archer and Don Tollefson. Tollefson was a popular sportscaster at WPVI before moving to Fox 29. (National Academy of Television Arts and Sciences.)

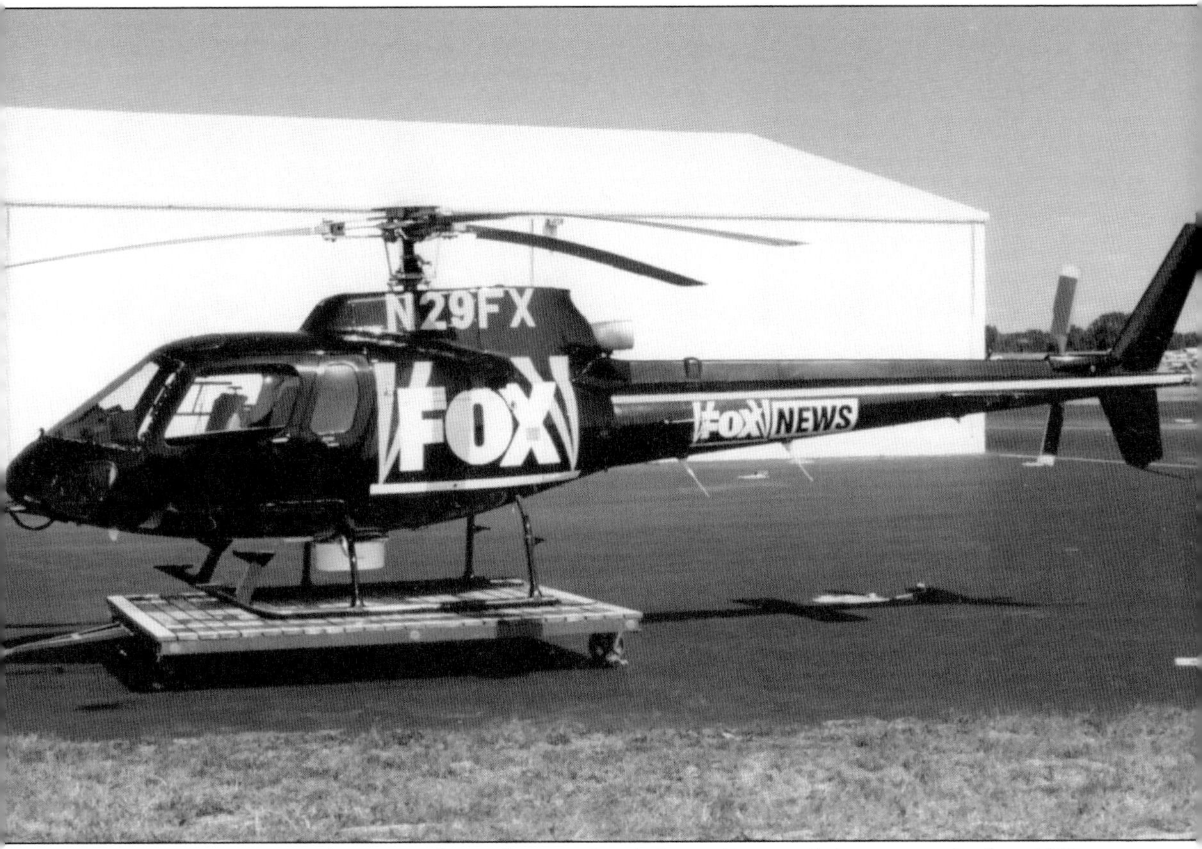

Fox Chopper 29 is seen here at its home base at South Jersey Regional Airport in the late 1990s. (Author's collection.)

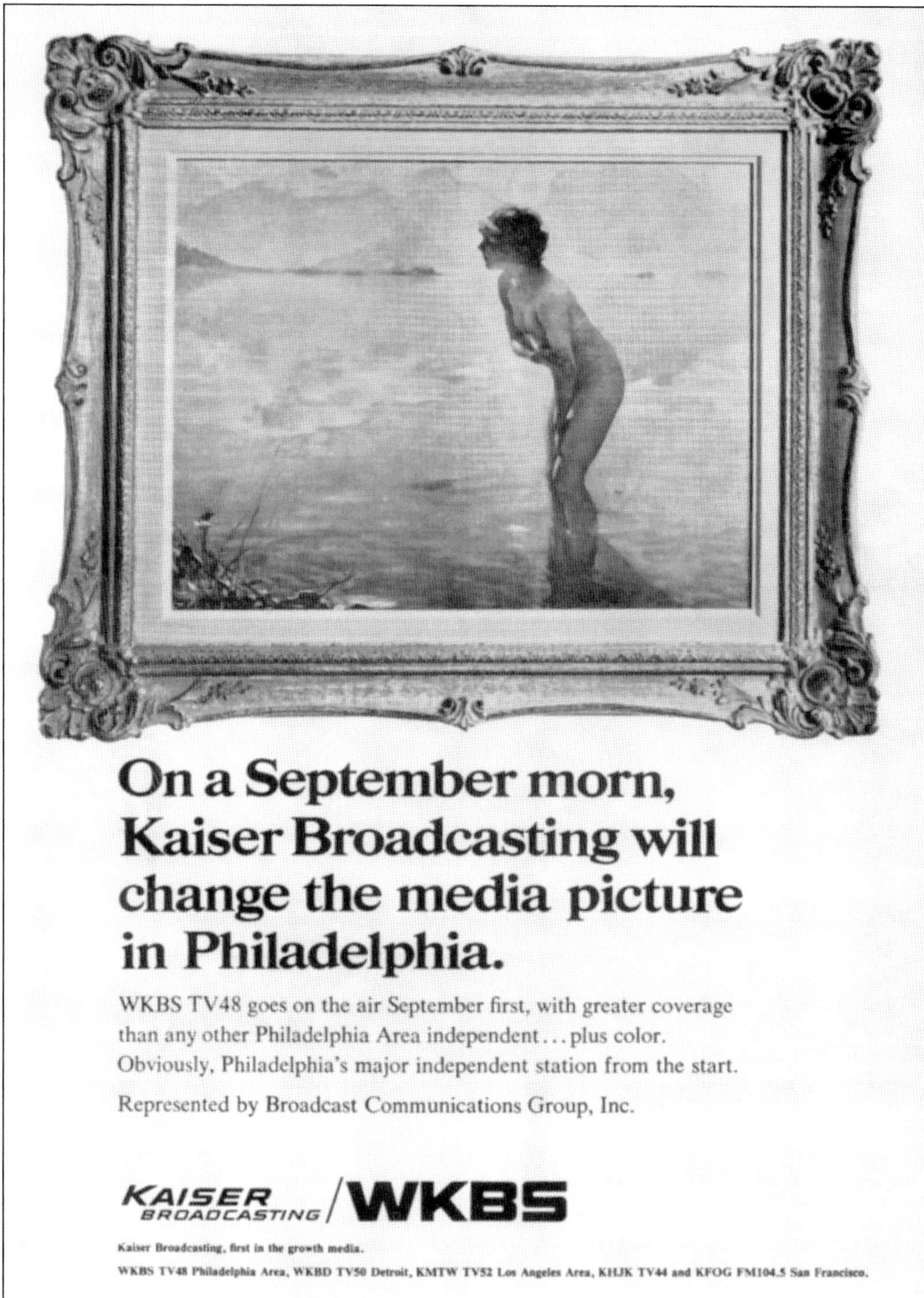

This is an advertisement for the debut of WKBS-TV Channel 48, which hit the Philadelphia airwaves on September 1, 1965. (Jay Lloyd.)

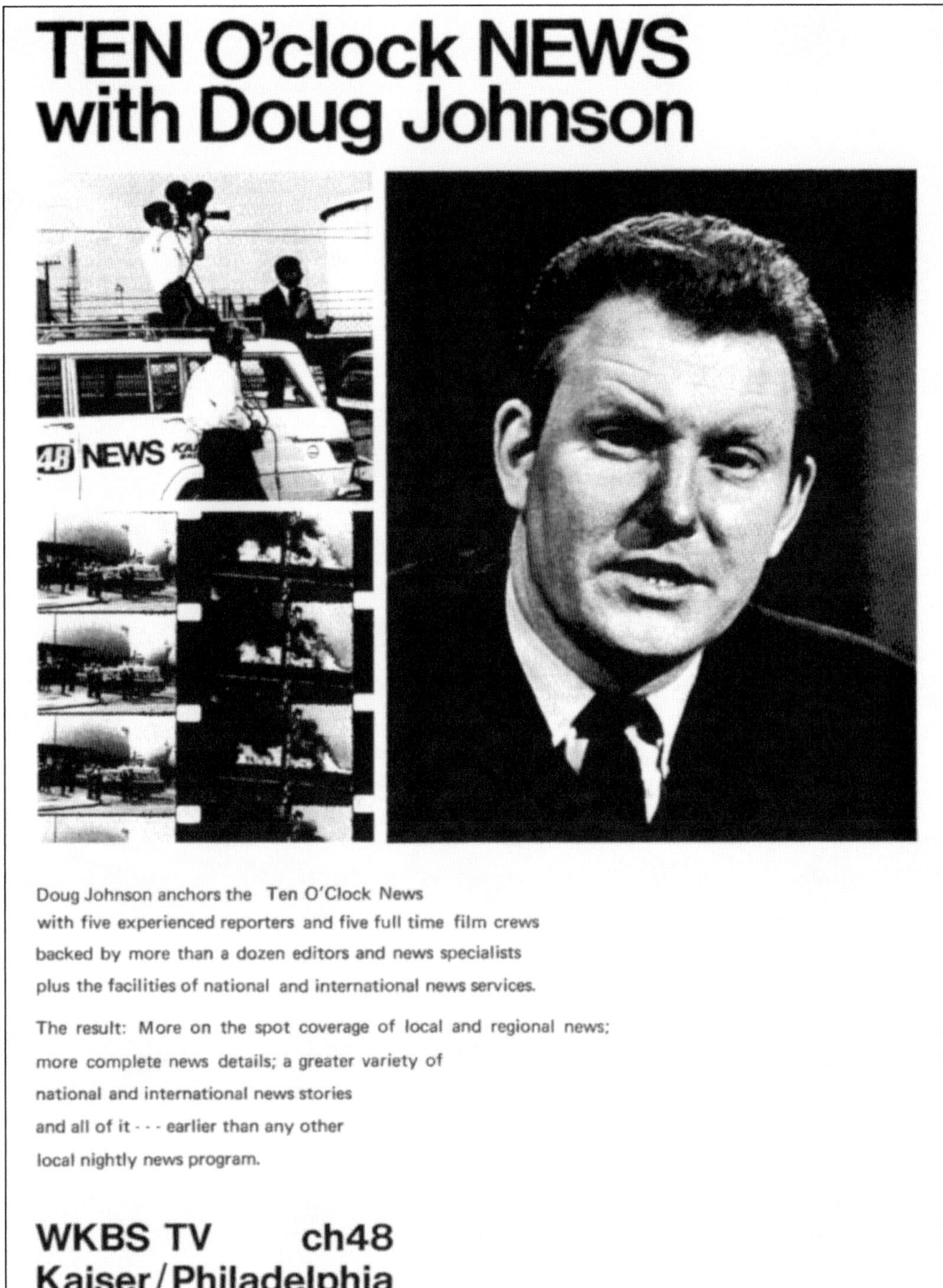

Channel 48 started the city's first 10 o'clock newscast in 1968, anchored by Doug Johnson until 1969. Johnson would later report for WABC-TV in New York. Longtime NBC4 Washington anchor Jim Vance also got his start at Channel 48. Pat Farnack anchored evening news updates from the late 1970s until the station went dark in 1983. (Jay Lloyd.)

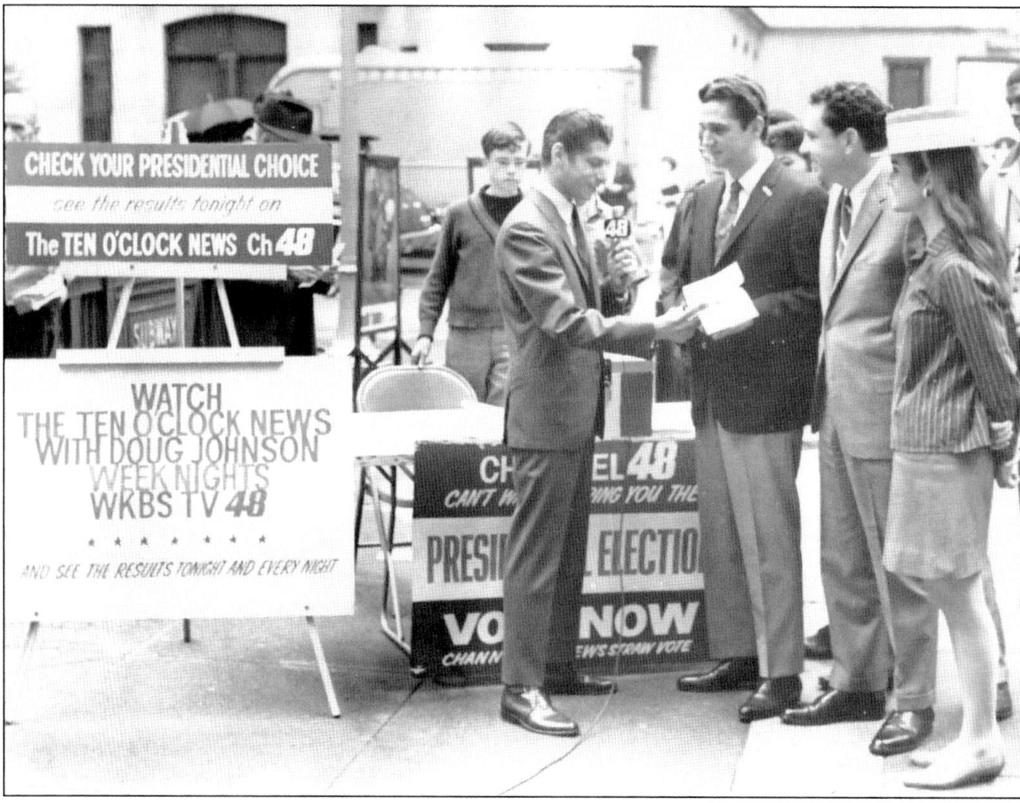

Jay Lloyd (holding the Channel 48 microphone) was a reporter for the fledgling *Ten O'Clock News*. Lloyd is seen here promoting the station's election night coverage outside of Philadelphia City Hall in 1968. (Jay Lloyd.)

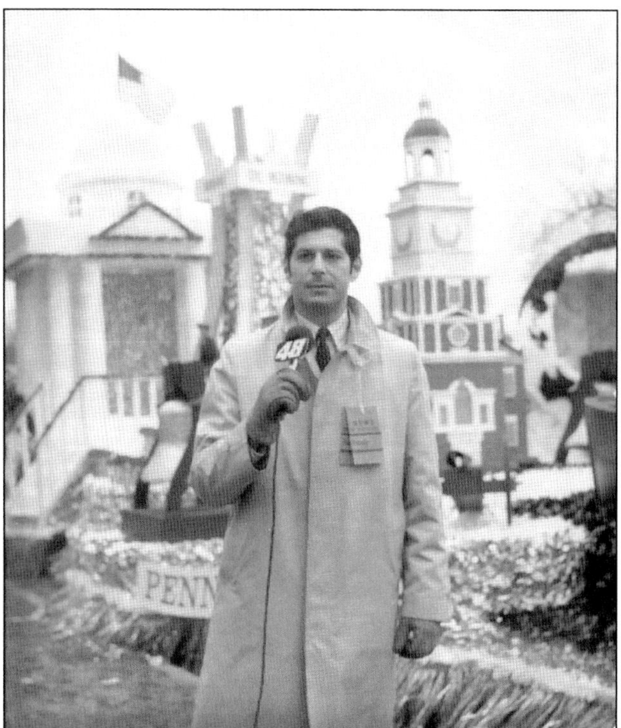

Lloyd reports on Pres. Richard Nixon's inauguration in January 1969. Lloyd was KYW Newsradio's suburban bureau chief from 1972 until his retirement in 2005. (Jay Lloyd.)

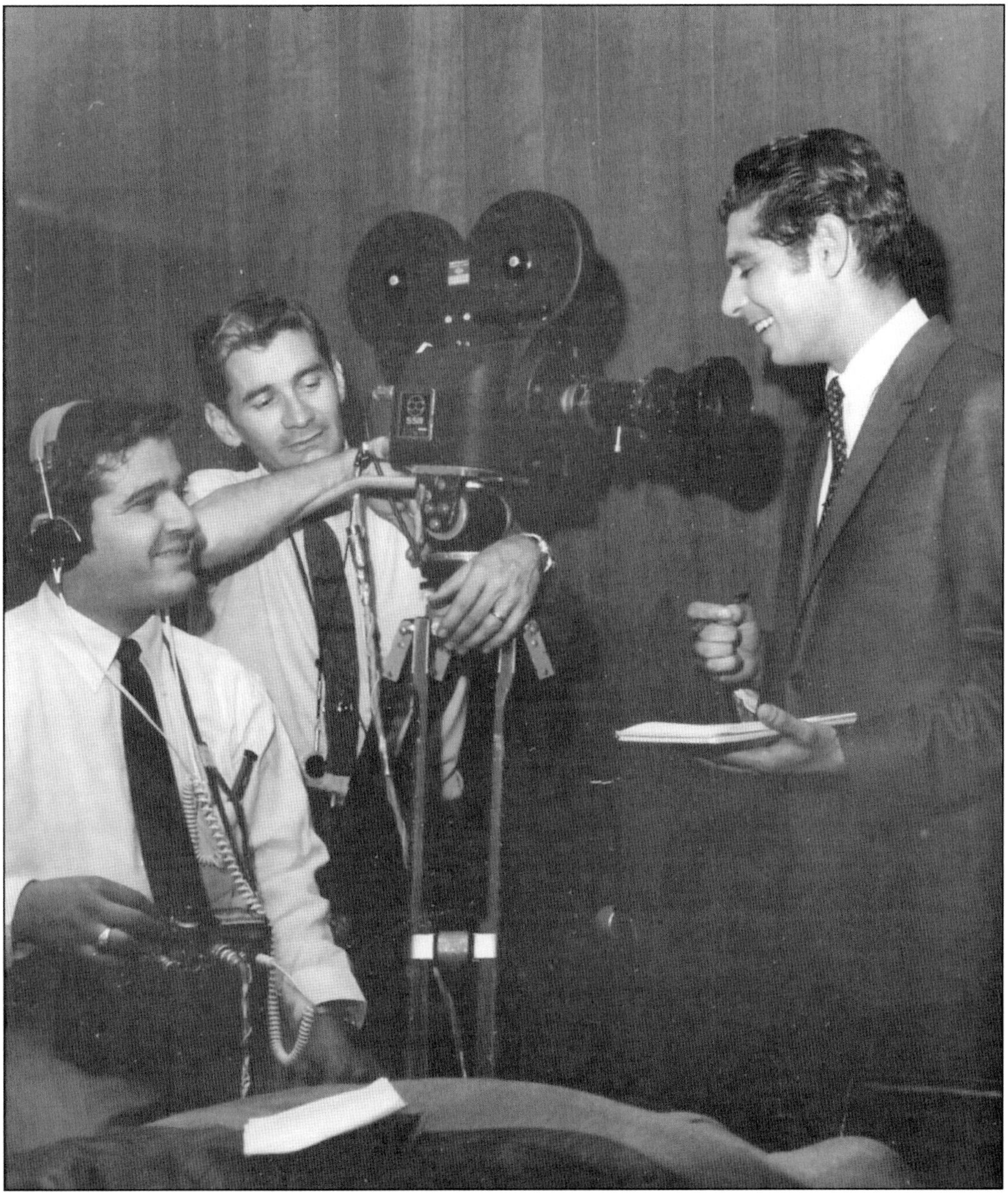

Here is Lloyd with his cameraman Don Spause and soundman Irv Grodsky in the early days of Channel 48's *Ten O'Clock News*. Grodsky went on to KYW, where he worked with Mike Strug in Car 3. (Jay Lloyd.)

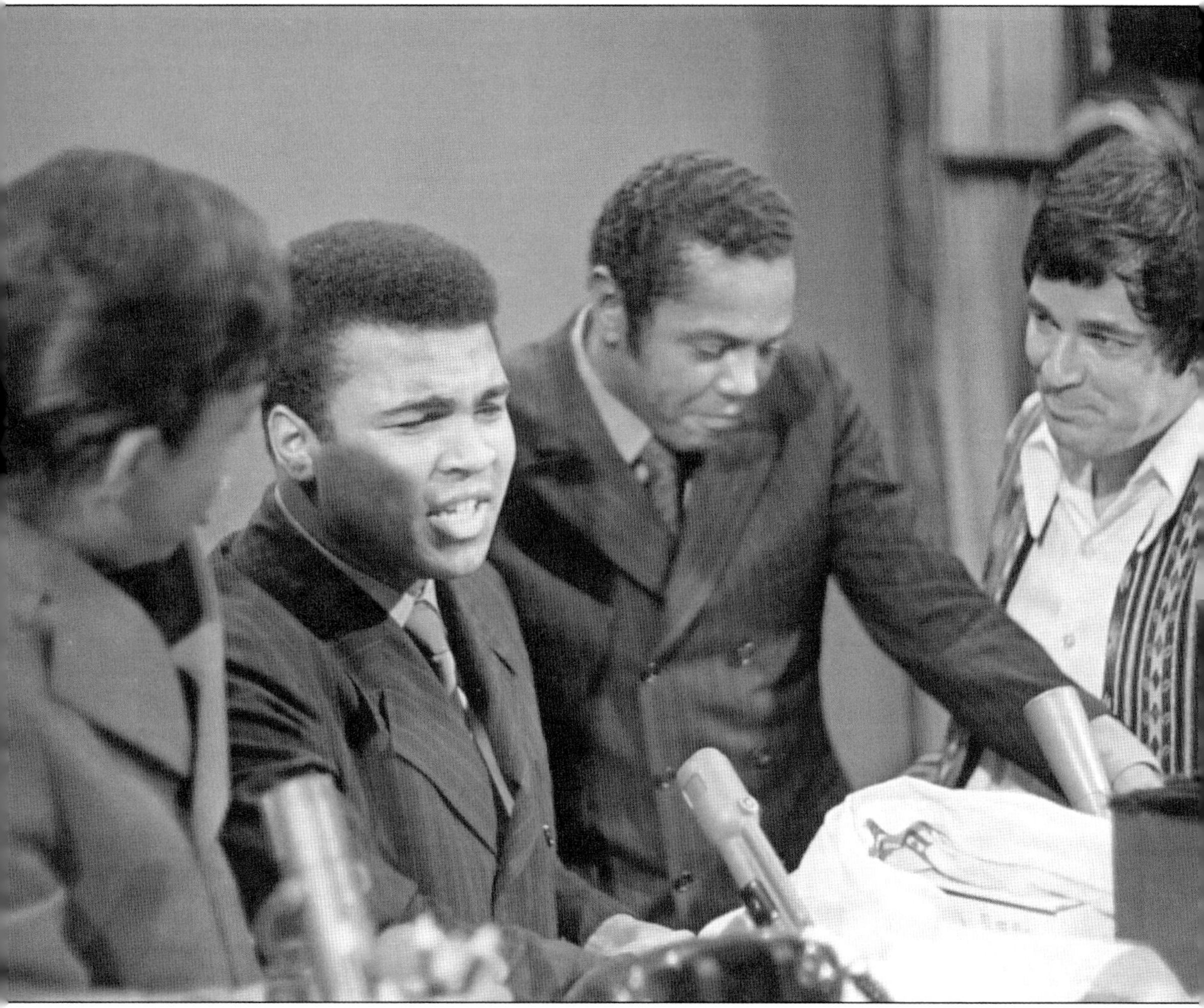

Like the other UHF stations, Channel 48 aired several telethons. In November 1969, boxing champion Muhammad Ali was a special guest on the *YAA Black Community Telethon*. From left to right are WDAS disc jockey Jimmy Bishop, Ali, WDAS disc jockey Georgie Woods, and WIBG disc jockey Hy Lit. Lit hosted the nationally syndicated *Hy Lit Show* at WKBS-TV. (Special Collections Research Center, Temple University Libraries, Philadelphia, Pennsylvania.)

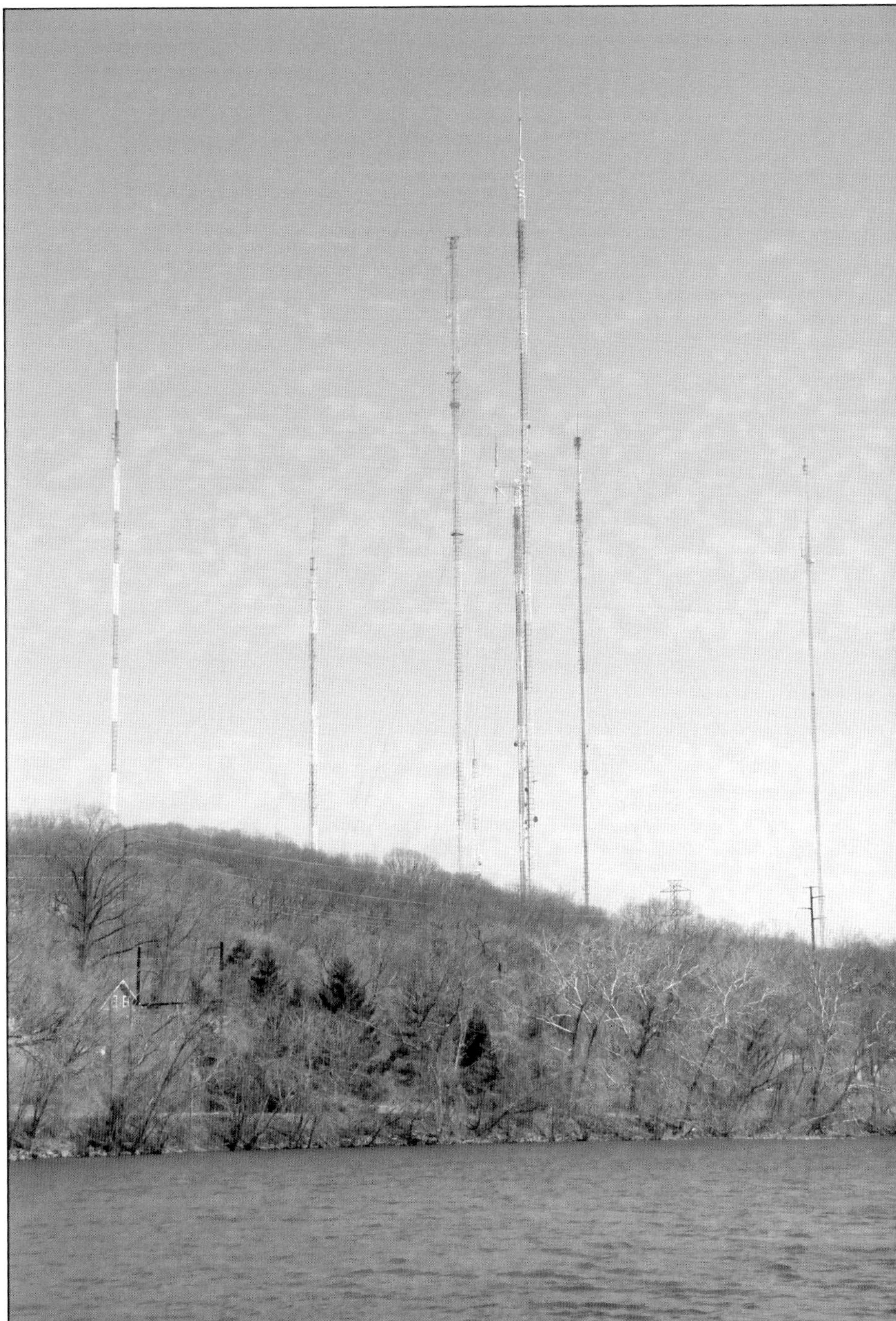

All of Philadelphia's UHF, VHF, and most radio stations broadcast from the Roxborough towers. As of 2015, there were eight towers over 1,000 feet tall (the tallest being Channel 3 and Channel 6's shared tower, which is 1,276 feet) and several shorter towers. They are located in the city's Roxborough neighborhood, one of the highest points in the city. The cluster of towers is a landmark that can be seen for miles. (Author's collection.)